Michael J. Sailor

Porous Silicon in Practice

The Author

Prof. Dr. Michael J. Sailor
University of California Mail
C 0358
Chemistry and Biochemistry
9500 Gilman Drive
La Jolla, CA 92093-0358
USA

All books published by **Wiley-VCH** are carefully produced. Nevertheless, authors, editors, and publisher do not warrant the information contained in these books, including this book, to be free of errors. Readers are advised to keep in mind that statements, data, illustrations, procedural details or other items may inadvertently be inaccurate.

Library of Congress Card No.: applied for

British Library Cataloguing-in-Publication Data
A catalogue record for this book is available from the British Library.

Bibliographic information published by the Deutsche Nationalbibliothek
The Deutsche Nationalbibliothek lists this publication in the Deutsche Nationalbibliografie; detailed bibliographic data are available on the Internet at <http://dnb.d-nb.de>.

© 2012 Wiley-VCH Verlag & Co. KGaA, Boschstr. 12, 69469 Weinheim, Germany

All rights reserved (including those of translation into other languages). No part of this book may be reproduced in any form – by photoprinting, microfilm, or any other means – nor transmitted or translated into a machine language without written permission from the publishers. Registered names, trademarks, etc. used in this book, even when not specifically marked as such, are not to be considered unprotected by law.

Cover Design Grafik-Design Schulz, Fußgönheim
Typesetting Toppan Best-set Premedia Limited, Hong Kong
Printing and Binding Fabulous Printers Pte Ltd, Singapore

Printed in Singapore
Printed on acid-free paper

Print ISBN: 978-3-527-31378-5
ePDF ISBN: 978-3-527-64192-5
oBook ISBN: 978-3-527-64190-1
ePub ISBN: 978-3-527-64191-8
Mobi ISBN: 978-3-527-64193-2

Contents

Preface *XI*

1	**Fundamentals of Porous Silicon Preparation** *1*	
1.1	Introduction *1*	
1.2	Chemical Reactions Governing the Dissolution of Silicon *2*	
1.2.1	Silicon Oxides and Their Dissolution in HF *3*	
1.2.2	Silicon Oxides and Their Dissolution in Basic Media *3*	
1.2.3	Silicon Hydrides *4*	
1.3	Experimental Set-up and Terminology for Electrochemical Etching of Porous Silicon *5*	
1.3.1	Two-Electrode Cell *6*	
1.3.2	Three-Electrode Cell *6*	
1.4	Electrochemical Reactions in the Silicon System *7*	
1.4.1	Four-Electron Electrochemical Oxidation of Silicon *8*	
1.4.2	Two-Electron Electrochemical Oxidation of Silicon *9*	
1.4.3	Electropolishing *10*	
1.5	Density, Porosity, and Pore Size Definitions *11*	
1.6	Mechanisms of Electrochemical Dissolution and Pore Formation *13*	
1.6.1	Chemical Factors Controlling the Electrochemical Etch *16*	
1.6.2	Crystal Face Selectivity *18*	
1.6.3	Physical Factors Controlling the Electrochemical Etch *18*	
1.7	Resume of the Properties of Crystalline Silicon *19*	
1.7.1	Orientation *19*	
1.7.2	Band Structure *20*	
1.7.3	Electrons and Holes *21*	
1.7.4	Photoexcitation of Semiconductors *22*	
1.7.5	Dopants *23*	
1.7.6	Conductivity *24*	
1.7.7	Evolution of Energy Bands upon Immersion in an Electrolyte *24*	
1.7.8	Charge Transport at p-Type Si Liquid Junctions *26*	

1.7.9	Idealized Current–Voltage Curve at p-Type Liquid Junctions 26
1.7.10	Energetics at n-Type Si Liquid Junctions 28
1.7.11	Idealized Current–Voltage Curve at n-type Liquid Junctions 28
1.8	Choosing, Characterizing, and Preparing a Silicon Wafer 28
1.8.1	Measurement of Wafer Resistivity 29
1.8.2	Cleaving a Silicon Wafer 34
1.8.3	Determination of Carrier Type by the Hot-Probe Method 36
1.8.4	Ohmic Contacts 36
1.8.4.1	Making an Ohmic Contact by Metal Evaporation 39
1.8.4.2	Making an Ohmic Contact by Mechanical Abrasion 40
	References 40

2 Preparation of Micro-, Meso-, and Macro-Porous Silicon Layers 43

2.1	Etch Cell: Materials and Construction 43
2.2	Power Supply 44
2.3	Other Supplies 48
2.4	Safety Precautions and Handling of Waste 48
2.5	Preparing HF Electrolyte Solutions 50
2.6	Cleaning Wafers Prior to Etching 51
2.6.1	No Precleaning 51
2.6.2	Ultrasonic Cleaning 51
2.6.3	RCA Cleaning 52
2.6.4	Removal of a Sacrificial Porous Layer with Strong Base 52
2.7	Preparation of Microporous Silicon from a p-Type Wafer 53
2.8	Preparation of Mesoporous Silicon from a p^{++}-Type Wafer 57
2.9	Preparation of Macroporous, Luminescent Porous Silicon from an n-Type Wafer (Frontside Illumination) 59
2.9.1	Power Supply Limitations 63
2.10	Preparation of Macroporous, Luminescent Porous Silicon from an n-Type Wafer (Back Side Illumination) 64
2.11	Preparation of Porous Silicon by Stain Etching 68
2.12	Preparation of Silicon Nanowire Arrays by Metal-Assisted Etching 73
	References 75

3 Preparation of Spatially Modulated Porous Silicon Layers 77

3.1	Time-Programmable Current Source 78
3.1.1	Time Resolution Issues 79
3.1.2	Etching with an Analog Source 80
3.1.3	Etching with a Digital Source 82
3.2	Pore Modulation in the z-Direction: Double Layer 83
3.3	Pore Modulation in the z-Direction: Rugate Filter 83
3.3.1	Tunability of the Rugate Spectral Peak Wavelength 88
3.3.2	Width of the Spectral Band 92

3.4	More Complicated Photonic Devices: Bragg Stacks, Microcavities, and Multi-Line Spectral Filters	*94*
3.4.1	Bragg Reflector	*96*
3.4.2	Multiple Spectral Peaks-"Spectral Barcodes"	*100*
3.5	Lateral Pore Gradients (in the x–y Plane)	*104*
3.6	Patterning in the x–y Plane Using Physical or Virtual Masks	*108*
3.6.1	Physical Masking Using Photoresists	*109*
3.6.2	Virtual Masking Using Photoelectrochemistry	*112*
3.7	Other Patterning Methods	*114*
	References	*114*

4 Freestanding Porous Silicon Films and Particles *119*

4.1	Freestanding Films of Porous Silicon-"Lift-offs"	*120*
4.2	Micron-scale Particles of Porous Silicon by Ultrasonication of Lift-off Films	*120*
4.3	Core–Shell (Si/SiO$_2$) Nanoparticles of Luminescent Porous Silicon by Ultrasonication	*126*
	References	*130*

5 Characterization of Porous Silicon *133*

5.1	Gravimetric Determination of Porosity and Thickness	*134*
5.1.1	Errors and Limitations of the Gravimetric Method	*137*
5.2	Electron Microscopy and Scanned Probe Imaging Methods	*138*
5.2.1	Cross-Sectional Imaging	*138*
5.2.2	Plan-View (Top-Down) Imaging	*139*
5.3	Optical Reflectance Measurements	*139*
5.3.1	Instrumentation to Collect Reflectance Data	*139*
5.3.1.1	Reflectance Optics	*140*
5.3.1.2	Wavelength Calibration	*142*
5.3.2	Principles of Fabry–Pérot Interference	*143*
5.3.3	Analyzing Fabry–Pérot Interference Spectra by Fourier Transform: the RIFTS Method	*150*
5.3.3.1	Preparation of Spectrum for Fast Fourier Transform	*151*
5.3.3.2	Interpretation of the Fast Fourier Transform	*153*
5.3.4	Thickness and Porosity by the Spectroscopic Liquid Infiltration Method (SLIM)	*154*
5.3.4.1	Bruggeman Effective Medium Approximation	*155*
5.3.4.2	Determination of Thickness and Porosity by SLIM	*156*
5.3.4.3	Determination of Index of Refraction of the Porous Skeleton	*156*
5.3.4.4	Effect of Skeleton Index on Porosity Determined by SLIM	*158*
5.3.5	Comparison of Gravimetric Measurement with SLIM for Porosity and Thickness Determination	*159*
5.3.6	Analysis of Double-Layer Structures Using RIFTS	*162*

5.4	Porosity, Pore size, and Pore Size Distribution by Nitrogen Adsorption Analysis (BET, BJH, and BdB Methods) *167*
5.5	Measurement of Steady-State Photoluminescence Spectra *170*
5.5.1	Origin of Photoluminescence from Porous Silicon *170*
5.5.1.1	Tunability of the Photoluminescence Spectrum *171*
5.5.1.2	Mechanisms of Photoluminescence *171*
5.5.2	Instrumentation to Acquire Steady-State Photoluminescence Spectra *173*
5.6	Time-Resolved Photoluminescence Spectra *173*
5.6.1	Long, Nonexponential Excited State Lifetimes *173*
5.6.2	Influence of Surface Traps *175*
5.7	Infrared Spectroscopy of Porous Silicon *176*
5.7.1	Characteristic Group Frequencies for Porous Silicon *176*
5.7.2	Measurement of FTIR Spectra of Porous Silicon *178*
5.7.2.1	Transmission Mode Measurement Using the Standard Etch Cell *179*
	References *181*
6	**Chemistry of Porous Silicon** *189*
6.1	Oxide-Forming Reactions of Porous Silicon *190*
6.1.1	Temperature Dependence of Oxidation Using Gas-Phase Oxidants *190*
6.1.2	Thermal (Air) Oxidation *191*
6.1.3	Ozone Oxidation *192*
6.1.4	High-Pressure Water Vapor Annealing *193*
6.1.5	Oxidation in Aqueous Solutions *193*
6.1.5.1	Aqueous Oxidation Induced by Cationic Surfactants *194*
6.1.6	Electrochemical Oxidation in Aqueous Mineral Acids *194*
6.1.7	Oxidation by Organic Species: Ketones, Aldehydes, Quinones, and Dimethylsulfoxide *195*
6.1.8	Effect of Chemical Oxidation on Pore Morphology *196*
6.2	Biological Implications of the Aqueous Chemistry of Porous Silicon *198*
6.3	Formation of Silicon–Carbon Bonds *200*
6.3.1	Thermal Hydrosilylation to Produce Si–C Bonds *200*
6.3.2	Working with Air- and Water-Sensitive Compounds – Schlenk Line Manipulations *201*
6.3.3	Classification of Surface Chemistry by Contact Angle *203*
6.3.4	Microwave-Assisted Hydrosilylation to Produce Si–C Bonds *204*
6.3.5	Chemical or Electrochemical Grafting to Produce Si–C Bonds *206*
6.4	Thermal Carbonization Reactions *208*
6.4.1	Thermal Degradation of Acetylene to form "Hydrocarbonized" Porous Silicon *208*

6.4.2	Thermal Degradation of Polymers to Form "Carbonized" Porous Silicon	*209*
6.5	Conjugation of Biomolecules to Modified Porous Silicon	*211*
6.5.1	Carbodiimide Coupling Reagents	*211*
6.5.2	Attachment of PEG to Improve Biocompatibility	*212*
6.5.3	Biomodification of "Hydrocarbonized" Porous Silicon	*213*
6.5.4	Silanol-Based Coupling to Oxidized Porous Silicon Surfaces	*215*
6.6	Chemical Modification in Tandem with Etching	*217*
6.7	Metallization Reactions of Porous Silicon	*218*
	References	*219*

Appendix A1. Etch Cell Engineering Diagrams and Schematics *229*

Standard or Small Etch Cell-Complete *229*
Standard Etch Cell Top Piece *230*
Small Etch Cell Top Piece *231*
Etch Cell Base (for Either Standard or Small Etch Cell) *232*
Large Etch Cell-Complete *232*
Large Etch Cell Top Piece *233*
Large Etch Cell Base *233*

Appendix A2. Safety Precautions When Working with Hydrofluoric Acid *235*

Hydrofluoric Acid Hazards *235*
First Aid Measures for HF Contact *236*
Note to Physician *238*
HF Antidote Gel *239*
Further Reading *239*

Appendix A3. Gas Dosing Cell Engineering Diagrams and Schematics *241*

Gas Dosing Cell Top Piece *242*
Gas Dosing Cell Middle Piece *243*
Gas Dosing Cell Bottom Piece *244*

Index *245*

Preface

This book is written for the beginner – someone who has no prior training in the field. It began as a series of summer tutorial lectures that I gave to my research group to familiarize them with the preparation and characterization of porous silicon. I found that the traditional undergraduate chemistry, biochemistry, bioengineering, physics, or materials science curriculum does not prepare one to work with porous silicon – most of my students would come into the group with no understanding of the electrochemical methods needed to carry out its synthesis, little appreciation for the fundamental semiconductor physics, electronics, chemistry, and optics principles needed to exploit its properties, and a sizable fear of the hydrofluoric acid used in its preparation. The tutorials resulted from my frustration that the basic conceptual and experimental "tricks of the trade" were not being passed from one student to the next. My goal was to provide my students with all that I thought they needed to know to get started in their research projects and survive the grilling of their second year oral committee. I provided laboratory "homework" experiments to get the students comfortable with the equipment and the techniques we use. The experiments in Chapters 1–5 are a direct result of these homework assignments. They are structured, step-by-step procedures with well-characterized results. I wrote them to allow me to correct obvious errors in laboratory technique or understanding before the student embarked on his or her independent research project, where errors are not as easily caught and carry significant consequences. The large increase in interest in porous silicon in the past few years, and the numerous email messages I have been receiving from students in groups around the world, asking me for details of our synthetic and optical analysis methods, gives me hope that more than my own research group members will make use of this material.

In the summer of 2004 I was fortunate to meet Esther Levy from Wiley-VCH, who, along with Martin Ottmar, encouraged me to convert my tutorial into a book. I thank them and the rest of the publishing team at Wiley-VCH for their patience during the several years spanning the writing and production of this work.

Many of my coworkers and collaborators contributed the ideas, concepts, and images that make up a large part of this book. In particular, I thank Gordon M. Miskelly, Giuseppe Barillaro, Andrea Potocny, Manuel Orosco, Sophia Oller, Ester Segal, M. Shaker Salem, Yukio H. Ogata, Stephanie Pace, Frederique Cunin, Jean-Marie Devoiselle, Luo Gu, Joseph Lai, Emily Anglin, Beniamino Sciacca, Michelle Y. Chen, Sara Alvarez, Anne M. Ruminski, Adrian Garcia Sega, and Vinh Diep.

Finally, I thank my family for putting up with the late nights, early mornings, and missed dinner appointments they suffered as I went through this process.

La Jolla
August 2011

Michael J. Sailor

1
Fundamentals of Porous Silicon Preparation

1.1
Introduction

Porous silicon was accidentally discovered by the Uhlirs, a husband and wife team working at Bell Laboratories in the mid 1950s. They were trying to develop an electrochemical method to machine silicon wafers for use in microelectronic circuits. Under the appropriate electrochemical conditions, the silicon wafer did not dissolve uniformly as expected, but instead fine holes appeared, propagating primarily in the <100> direction in the wafer. Since this did not provide the smooth polish desired, the curious result was reported in a Bell labs technical note [1], and then the material was more or less forgotten. In the 1970s and 1980s a significant level of interest arose because the high surface area of porous silicon was found to be useful as a model of the crystalline silicon surface in spectroscopic studies [2–5], as a precursor to generate thick oxide layers on silicon, and as a dielectric layer in capacitance-based chemical sensors [6].

Interest in porous silicon, and in particular in its nanostructure, exploded in the early 1990s when Ulrich Goesele at Duke University identified quantum confinement effects in the absorption spectrum of porous silicon, and almost simultaneously Leigh Canham at the Defense Research Agency in England reported efficient, bright red–orange photoluminescence from the material [7, 8]. The quantum confinement effects arise when the pores become extensive enough to overlap with each other, generating nanometer-scale silicon filaments. As expected from the quantum confinement relationship [9], the red to green color of photoluminescence occurs at energies that are significantly larger than the bandgap energy of bulk silicon (1.1 eV, in the near-infrared).

With the discovery of efficient visible light emission from porous silicon came a flood of work focused on creating silicon-based optoelectronic switches, displays, and lasers. Problems with the material's chemical and mechanical stability, and its disappointingly low electroluminescence efficiency led to a waning of interest by the mid 1990s. In the same time period, the unique features of the material – large surface area, controllable pore sizes, convenient surface chemistry, and compatibility with conventional silicon microfabrication technologies – inspired research into applications far outside optoelectronics. Many of the fundamental chemical stability problems have been overcome as the chemistry of the material has matured, and various biomedical [10–18] sensor, optics, and electronics applications have emerged [10].

Porous silicon is generated by etching crystalline silicon in aqueous or non-aqueous electrolytes containing hydrofluoric acid (HF). This book describes basic electrochemical and chemical etching experiments that can be used to make the main types and structures of porous silicon. Beginning with measurement of wafer resistivity, the experiments are intended for the newcomer to the field, written in the form of detailed procedures, including sources for the materials and equipment. Experiments describing methods for characterization and key chemical modification reactions are also provided. The present chapter gives an overview of fundamentals that are a useful starting point to understand the theory underlying the experiments in the later chapters.

1.2
Chemical Reactions Governing the Dissolution of Silicon

The formation of porous silicon involves reactions of Si–Si, Si–H, Si–O, and Si–F bonds at the surface of the silicon crystal. The relative strengths of these bonds, obtained from thermodynamic measurements of molecular analogues, are given in Table 1.1. While one might think that the strengths of these bonds would determine the relative stability of each species on a silicon surface, the electronegativity of the elements plays a much more important role. Si–H and Si–C species tend to passivate the silicon surface in aqueous solutions, while the Si–F bond is highly reactive. Electronegative elements such as O and F form more polar Si–X bonds, making the silicon

Table 1.1 Enthalpies of some Si–X bonds.

Compound	Bond	Enthalpy, kcal mol^{-1}
Me$_3$Si–SiMe$_3$	Si–Si	79
Me$_3$Si–CH$_3$	Si–C	94
Me$_3$Si–H	Si–H	95
Me$_3$Si–OMe$_3$	Si–O	123
Me$_3$Si–F	Si–F	158

Taken from Robin Walsh, Gelest Catalog: www.gelest.com

atom susceptible to nucleophilic attack. The surface of freshly prepared porous silicon is covered with a passivating layer of Si–H bonds, with minor quantities of Si–F and Si–O species.

1.2.1
Silicon Oxides and Their Dissolution in HF

Silicon is thermodynamically unstable in air or water, and it reacts spontaneously to form an oxide layer. The oxide can be nonstoichiometric and hydrated to various degrees, though the simple empirical formula is silicon dioxide, SiO_2 (Equation 1.1). SiO_2 is a key thermodynamic sink in the silicon system.

$$Si + O_2 \rightarrow SiO_2 \tag{1.1}$$

SiO_2 is an electrical insulator that forms passivating films on crystalline silicon; preparation of porous silicon thus requires an additive in the solution to dissolve the oxide and allow electrochemical oxidation to continue. The Si–F bond is the only bond stronger than Si–O, and it is the Si–F bond enthalpy that drives the main chemical dissolution reaction used to make porous silicon. In the presence of aqueous HF, SiO_2 spontaneously dissolves as SiF_6^{2-} (Equation 1.2).

$$SiO_2 + 6\,HF \rightarrow SiF_6^{2-} + 2\,H^+ + 2\,H_2O \tag{1.2}$$

The reaction of SiO_2 with HF is a common industrial reaction. It is used to prepare frosted glass from plate glass and to remove SiO_2 masking layers in the processing of silicon wafers in microelectronics. The silicon hexafluoride ion (SiF_6^{2-}) is a stable dianion that is highly soluble in water. Thus fluoride is the most important additive used in the preparation of porous silicon, dissolving the insulating oxide that would otherwise shut down the electrochemical corrosion reaction.

1.2.2
Silicon Oxides and Their Dissolution in Basic Media

In the absence of fluoride ion, SiO_2 on the surface of a silicon wafer protects the underlying silicon from further oxidation. While this is true in acidic or neutral aqueous solutions, in basic solutions hydroxide ions attack and dissolve the oxide by Equation 1.3:

$$SiO_2 + 2\,OH^- \rightarrow [SiO_2(OH)_2]^{2-} \tag{1.3}$$

The net dissolution reaction for silicon in basic media then becomes:

$$Si + 2\,OH^- + 2\,H_2O \rightarrow [SiO_2(OH)_2]^{2-} + 2\,H_2 \tag{1.4}$$

The reaction represented by Equation 1.3 is highly simplified. The species $[SiO_2(OH)_2]^{2-}$, the doubly protonated form of silicic acid, is only one of many water-soluble forms of silicon oxide. The fundamental oxide-containing unit is the SiO_4^{4-} tetrahedron, known as the orthosilicate ion [11]. Orthosilicate itself is highly basic, and in aqueous solutions it is never present as the naked SiO_4^{4-} ion. The fully protonated species is orthosilicic acid, $Si(OH)_4$, and this is the generic formula that is often presented in the literature to indicate all the water-soluble forms of silicic acid. The first ionization constant (pK_a) of $Si(OH)_4$ is about 10, and the second (pK_a) is around 12. Thus the $[SiO_2(OH)_2]^{2-}$ ion depicted in Equation 1.3 is only present in highly basic (pH > 12) solutions. In neutral or acidic solutions, $Si(OH)_4$ is the predominant monomeric form.

When the solution concentration of $Si(OH)_4$ is sufficiently large, silicic acid condenses into oligomers. Various "polysilicic acids" with the general formula $[SiO_x(OH)_{4-2x}]_n$, where $2 > x > 0$, are present in solution [11]. In neutral or acidic solutions these oligomers can condense to the point of precipitation, essentially the reverse of Equation 1.3:

$$Si(OH)_4 \rightarrow SiO_2 + 2\,H_2O \qquad (1.5)$$

The reaction represented in Equation 1.5 is the key reaction in the "sol–gel" process, often used to prepare colloids, films, or monoliths of porous silica from solution precursors [12]. The insolubility of SiO_2 in acidic solutions explains why elemental silicon does not corrode appreciably at pH < 7; the oxide provides a protective, passivating layer. The same is not true in basic solutions; here the solubility of silicon oxide drives silicon oxidation and dissolution by Equation 1.4. The high surface area and relatively strained nature of Si–Si bonds in porous silicon make the reaction with hydroxide ion quite rapid. In Chapter 2 we will employ this reaction (using aqueous KOH) to dissolve a porous silicon layer in order to determine its porosity. The Si–Si bonding in bulk silicon is less strained, and bulk silicon dissolves more slowly in basic solutions. In these and other situations where the oxide is soluble, dissolution of silicon becomes limited by surface Si–H species.

1.2.3
Silicon Hydrides

The reaction of silicon with water should be analogous to the reaction of metallic sodium in water; elemental silicon is electropositive enough to spontaneously liberate hydrogen from water. However, silicon does not dissolve in acidic solutions, even if the solution contains fluoride ion to remove the passivating SiO_2 layer. Although thermodynamically feasible, dissolution of silicon in aqueous HF is slow unless strong oxidizing agents (such as O_2 or NO_3^-) are present in the solution, or unless the oxidation reaction is driven by electrochemistry. The reason is that corrosion becomes kinetically limited by the passivating nature of surface hydrides.

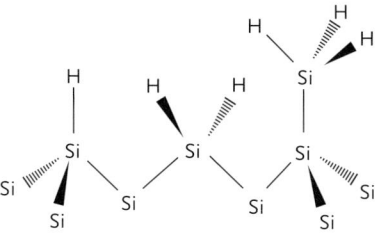

Figure 1.1
Hydrides on the porous silicon surface. The freshly etched surface of porous silicon is terminated primarily with hydride species. Residual oxides or fluorides are removed by the HF electrolyte. Three types of surface hydrides are depicted: SiH, SiH_2, and SiH_3.

When silicon is chemically or electrochemically etched in HF-containing solutions, the exposed silicon surface becomes terminated with H atoms (Figure 1.1). The mechanism of this reaction is described in more detail later in this chapter. The surface Si–H species are not readily removed by acid, and they must be oxidized to allow the silicon corrosion reaction to continue. In alkaline solutions, OH^- is able to attack these species because it is a good nucleophile. Nucleophilic attack is an important reaction in the silicon system, and it is discussed in more detail in Chapter 6. The Si–H species on porous silicon can also be removed by the action of a cationic surfactant, which polarizes the surface and induces nucleophilic attack by water, even in acidic solutions [13]. This reaction is also discussed in Chapter 6.

1.3
Experimental Set-up and Terminology for Electrochemical Etching of Porous Silicon

In an electrochemical reaction, two electrodes are needed. One supplies electrons to the solution (the cathode) and the other removes electrons from the solution (the anode). It is important to keep in mind that the two electrodes are required to maintain charge neutrality and to complete the electrical circuit. Regardless of the oxidation or reduction reactions occurring at the electrodes, you cannot perform electrochemistry if you do not complete the circuit. This means that at least two reactions are occurring simultaneously in an electrochemical cell, the anode (oxidation) reaction and the cathode (reduction) reaction. Electrochemists refer to these as "half-reactions." A schematic of a two-electrode cell for etching silicon, with the relevant half-reactions, is shown in Figure 1.2.

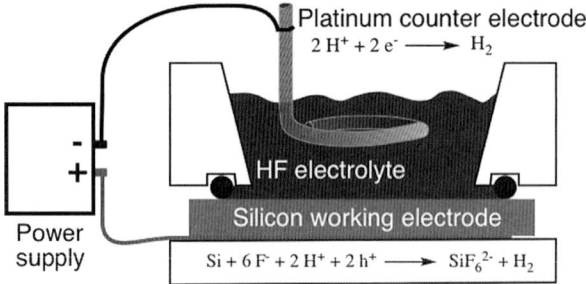

Figure 1.2
Schematic of a two-electrode electrochemical cell used to make porous silicon. Silicon is the working electrode. The working electrode is an anode in this case, because an oxidation reaction occurs at its surface. The cathode counter-electrode is typically platinum. The main oxidation and reduction half-reactions occurring during the formation of porous silicon are given.

1.3.1
Two-Electrode Cell

In the two-electrode cell, electrochemical reactions must occur at both electrodes, but generally you are interested in the reaction at only one of these. In the case of porous silicon formation, the silicon electrode is the important one. It is the anode, and the chemical being oxidized is the silicon itself. The cathode used in porous silicon etching cells is usually platinum, and it is separated from the silicon electrode by a few mm to several cm of electrolyte solution, or in some cases by a membrane or salt bridge. The electrochemical reaction occurring at the platinum electrode is primarily the reduction of protons to hydrogen gas. Electrochemists refer to the silicon electrode as the "working electrode", and the platinum electrode as the "counter-electrode" in this experiment. We are generally not concerned with the counter-electrode (cathode) reaction, although it can produce byproducts that interfere with the silicon electrocorrosion reaction or otherwise limit the silicon etching process. The design of electrochemical cells and practical considerations regarding the counter-electrode and other cell materials are discussed in Chapter 2.

1.3.2
Three-Electrode Cell

A three-electrode electrochemical cell is used when one wants to measure both the current and the potential of an electrochemical reaction simultaneously [14]. This situation is commonly encountered in electrochemistry, because it allows the identification of the driving force of an electrochemical process. In a three-electrode configuration, a high-impedance reference electrode is connected into a specialized feedback circuit. The reference electrode is designed to precisely control the electrochemical processes that can occur at its surface, so that the electrochemical potential is well defined and does not drift. A common reference electrode used in aqueous systems is the saturated calomel electrode. It is defined by the half-reaction of mercury metal (Hg) with mercury (I) chloride (Hg_2Cl_2) in a solution saturated with potassium chloride (KCl). The mercury, mercury (I) chloride, and potassium chloride are all contained in a small tube separated from the electrochemical cell by a porous glass membrane (typically Vycor glass). The presence of the membrane and the high impedance (resistance to current flow) established by the measurement circuitry ensure that only a small ion current is allowed to flow between the reference electrode compartment and the rest of the electrochemical cell, maintaining a stable reference potential. This sort of cell is not compatible with solutions containing HF, and so various other reference electrode configurations have been used for porous silicon experiments.

A common reference electrode for HF systems is a bare platinum wire placed in a solution of HF, separated from the main electrochemical compartment by a thin plastic capillary tube. This is not a particularly reproducible reference electrode, since the electrochemical reaction potentials are sensitive to surface impurities on the Pt wire. Nevertheless, the electrode potential is fairly stable over periods of hours, and it is common practice to use such electrodes, referred to as "pseudo-reference" electrodes to indicate their tenuous relationship to true thermodynamic potentials. Throughout this book we are less concerned with the potential at the silicon surface and more concerned with the total current flowing through it. This is because most of the key properties of a porous silicon film: the porosity, pore size, and thickness are determined by the current. A two-electrode configuration is sufficient to set this parameter, and it is used in the experiments described in this book.

1.4
Electrochemical Reactions in the Silicon System

A representative current–potential curve for silicon in an HF electrolyte is shown in Figure 1.3. The curve displays characteristics common to the Si/HF system: an initial exponential rise in current with applied potential that reaches a maximum, decreases somewhat, and then increases more slowly at increasingly positive potentials. There are three regions usually defined: the porous Si formation region, a transition region, and the electropolishing regime.

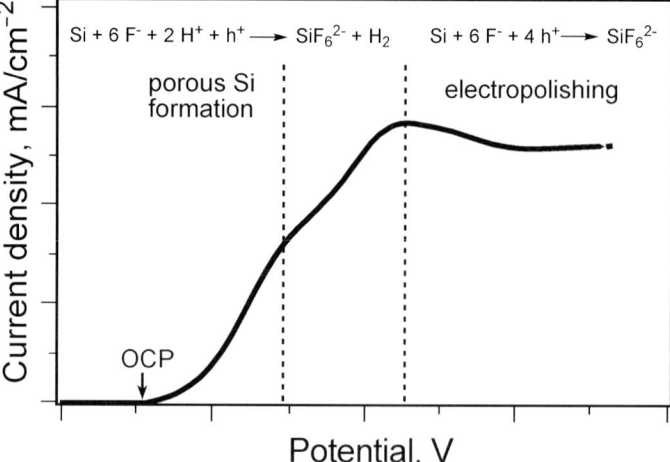

Figure 1.3
General current density versus applied potential curve for electrochemical etch of silicon in an HF electrolyte, showing the regimes for porous silicon formation and for electropolishing. The relevant 2-electron and 4- electron oxidation reactions are shown. "OCP" indicates the open circuit potential, or rest potential, of the silicon electrode. Data for this curve correspond to a moderately doped p-type silicon wafer in 1% HF solution, adopted from Zhang, X. G. Morphology and Formation Mechanisms of Porous Silicon. *J. Electrochem. Soc.* 151, C69–C80 (2004).

1.4.1
Four-Electron Electrochemical Oxidation of Silicon

To drive the corrosion of silicon electrochemically, positive current must be passed through the silicon electrode. The simplest reaction that can be expected during anodic dissolution of silicon in fluoride-containing solutions is the 4-electron oxidation represented by Equation 1.6. Here we use holes in the silicon valence band as the oxidizing equivalents. Note this equation is written as a half-reaction. As mentioned above, the electrons supplied by silicon at the working electrode must be balanced by a reduction half-reaction that consumes electrons at the counter-electrode. The electrochemical process performed by these electrons is usually the reduction of water to hydrogen gas.

$$\text{Anode (working electrode)}: Si + 6\ F^- + 4\ h^+ \rightarrow SiF_6^{2-} \tag{1.6}$$

It turns out that the 4-electron half-reaction represented by Equation 1.6, the anodic oxidation of silicon, is the predominant reaction when the electrocorrosion reaction is running at "full speed," and no porous silicon is being formed. This condition is referred to as electropolishing, and it occurs at the more positive electrode potentials depicted in Figure 1.3. It should be noted that this is an idealized equation, and that other compounds are involved in the silicon-HF system during electrocorrosion. For example, SiF_2, SiF_4, and the various members of the $SiH_nF_{(4-n)}$ family are thermodynamically accessible, though they are generally minor byproducts in the reaction. The active fluoride-containing species in aqueous and non-aqueous HF solutions include HF, $(HF)_2$, and HF_2^-.

1.4.2
Two-Electron Electrochemical Oxidation of Silicon

When the Uhlirs first prepared porous silicon, they noted bubbles rising from the silicon wafer (Figure 1.4) [1]. They assumed this gas was oxygen. For a "normal" aqueous electrochemical process, splitting of water is an often-encountered side reaction – especially if the applied potential exceeds the thermodynamic water splitting potential of 1.23 V and the electron transfer kinetics of the reaction of interest are sluggish. In the electrolysis

Figure 1.4
Electrochemical etching of silicon in the current regime where porous silicon is formed. This is a top view of an electrochemical etching cell. Bubbles of hydrogen gas are observed forming at the silicon surface.

of water, hydrogen gas is expected to form at the cathode (platinum wire in our case) and oxygen gas should form at the anode (the silicon wafer). The bubbles rising from the platinum counter-electrode (cathode) are indeed hydrogen, coming from the water electrolysis reaction. When the Uhlirs collected the gas coming from the silicon electrode, they expected it was oxygen and performed the classical glowing wooden splint experiment: placing the burning embers of a wooden stick into pure oxygen causes the stick to catch fire. Instead of a flame, they observed a small but startling explosion – the gas they had collected was hydrogen.

The unexpected evolution of hydrogen during the electrocorrosion of silicon is related to the fact that spontaneous reduction of water by silicon is thermodynamically favored. The kinetics of this reaction are slow for silicon in its elemental form (oxidation state 0), but silicon in its +2 oxidation state reacts rapidly with water, liberating hydrogen and producing silicon in its most common +4 oxidation state. The 2-electron oxidation process is represented in the two-step formalism of Equations 1.7 and 1.8. As in Equation 1.6, we use holes in the silicon valence band as the oxidizing equivalents. Note the lower case h^+ depicts a valence band hole and the upper case H^+ is a proton in these equations.

$$\text{Electrochemical step: } Si + 2\ F^- + 2\ h^+ \rightarrow [SiF_2] \tag{1.7}$$

$$\text{Chemical step: } [SiF_2] + 4\ F^- + 2\ H^+ \rightarrow SiF_6^{2-} + H_2 \tag{1.8}$$

$$\text{Net: } Si + 6\ F^- + 2\ H^+ + 2\ h^+ \rightarrow SiF_6^{2-} + H_2 \tag{1.9}$$

The two-electron process of Equation 1.9 predominates at lower applied potentials, and it is the main half-reaction responsible for porous silicon formation. The region of the current density–potential plot in Figure 1.3 labeled "PS formation" corresponds to this regime.

1.4.3
Electropolishing

When the current–potential relationship is in the electropolishing regime, silicon atoms are removed isotropically (i.e., no pores form). The net result is that the silicon wafer becomes thinner. This was the goal that the Uhlirs were pursuing in the mid 1950s when they accidentally discovered porous silicon. Electropolishing usually follows the 4-electron dissolution stoichiometry of Equation 1.6. Although some roughening can occur, electropolishing generally removes silicon atoms uniformly, and a smooth, polished surface will result if a polished wafer is used at the outset. An electropolishing reaction can be used to remove a pre-formed porous silicon layer from the silicon substrate (a "lift-off," see Experiment 4.1). In this case the electropolishing reaction occurs at the porous silicon/crystalline silicon interface, undercutting the porous layer.

1.5
Density, Porosity, and Pore Size Definitions

Porous materials are less dense than the constituent materials from which they are made because they contain voids. These voids can be open to the outside world or closed off from it. The IUPAC recommendations for classifying porous media include definitions of density to account for the open or closed nature of the pores [15]:

Density:
true density: density of the material excluding pores and voids
bulk density: density of the material including pores, voids, and closed and inaccessible pores
apparent density: density of the material including closed and inaccessible pores

An illustrative example in the case of porous silicon: If we ignore the volume taken up by the surface hydrogen atoms, the "true density" of any porous silicon sample would be just the density of crystalline silicon found in the textbooks, $2.33\,\text{g}\,\text{ml}^{-1}$. To determine the "bulk density," we need to know the geometry of our sample. Assume we use the Standard etch cell (Appendix 1), and etch a disc of porous silicon 1.2 cm in diameter (d) and 50 µm thick (L). The volume of the porous disc would be $\pi(d/2)^2 L = 6.0 \times 10^{-3}\,\text{cm}^3$, or 6 microliters. If the sample is 75% porous, the disc alone (removed from the silicon wafer) weighs 3.49 mg. The bulk density would be the mass of the sample divided by the volume it occupies, or $3.49 \times 10^{-3}/6.0 \times 10^{-3} = 0.58\,\text{g}\,\text{ml}^{-1}$. Note that because the volume calculation includes the air in the voids, the bulk density is always smaller than the true density.

The value calculated for the apparent density is highly dependent on the method used to determine the volume of the sample. One common method to determine apparent volume is by liquid displacement. The sample is placed in a liquid of known volume, and the new volume is measured. The volume of the sample is just the difference in the two volumes. In this case, the fluid will only penetrate the pores that are accessible. In other words, the pores cannot be completely closed off, and they have to be large enough to accommodate the molecular dimensions of the liquid compound used. Thus the "apparent" density is as much a measure of the probe molecule as it is a measure of the sample.

The IUPAC defines porosity:

Porosity: ratio of the total pore volume V_p to the apparent volume V.
The meaning of the term "total pore volume" is dependent on the method. We must distinguish between "open porosity," "closed porosity", and "total porosity":

Open porosity: the volume of pores accessible to a given probe molecule

Closed porosity: the volume of pores that are inaccessible to the probe molecule

Total porosity: open porosity + closed porosity; the volume of pores accessible to a given probe molecule plus the volume of pores that are inaccessible to the probe

Since it is prepared by corrosion of solid crystalline silicon, one can safely assume that all the pores in an as-formed porous silicon sample had to be accessible to the corrosion solution at the time of formation. It is, therefore, an open porous material with no inaccessible voids. However, if the material has been annealed, such that some of the pore mouths close off, the material will contain closed porosity. Also, because of the volume increase associated with conversion of Si to SiO_2, it is possible that a porous silicon sample will develop closed, inaccessible pores as it oxidizes. A similar situation could derive from chemical derivatization of the inner pore walls of a porous silicon sample, as we shall see in Chapter 6. Finally, it is important to keep in mind that the term "open porosity" is defined relative to a given probe molecule; a small pore that accommodates a molecule like ethanol may not accept a larger protein molecule. This is particularly true in biosensor and drug delivery applications, where the molecules of interest tend to be large.

We define three different pore size regimes, based again on the IUPAC "Recommendations for the characterization of porous solids:" [15]:

Micropores have widths smaller than 2 nm.
Mesopores have widths between 2 and 50 nm.
Macropores have widths larger than 50 nm.

The terms "nanoporous," "nanopore," and so on have come into vogue in recent years. It would be meaningless to follow the SI convention for the Greek prefixes, which would require that a nanopore be 1000 times smaller than a micropore—significantly smaller than the diameter of a hydrogen atom! While it currently carries no officially accepted definition, general usage indicates that a "nanomaterial" has structural features of the order of 100 nm or less. Thus the term nanoporous can be considered to be a general descriptor referring to any of the above macro-, meso-, or microporous materials with pore sizes less than 100 nm. In fact, this is more in line with the original Greek meaning of nano, which translates as "dwarf."

Experiments 5.1 and 5.3 describe methods to measure porosity. Pore size can be measured by atomic force microscopy (AFM), high resolution scanning electron microscopy (SEM), or gas adsorption measurements. Measurement of adsorption isotherms of gases such as N_2 or CO_2 at low temperatures allows an indirect and widely accepted means of determination of pore size; the data are generally interpreted using models for gas adsorption, such as the BET, BJH, and BdB methods, which yield information on surface area and pore size, and a general idea of the shape of the pores. There are important assumptions related to pore shape and connectivity used in these models, and acquisition and interpretation of such data are beyond the scope of this book. A general description of the methods are given in Chapter 5; for more details the reader should consult the relevant literature [16, 17].

The collection of cryogenic nitrogen adsorption isotherm data is one example of an experiment that determines pore size based on admission of a probe molecule. The experiment measures gas pressure in a chamber that contains the sample of interest and a known amount of nitrogen at a low temperature. The pore size can also be inferred from an optical reflectivity experiment that monitors the refractive index of a film upon exposure to probe molecules of various sizes. For example, the admission or exclusion of proteins based on size has been determined using this method [18–21].

1.6
Mechanisms of Electrochemical Dissolution and Pore Formation

The process controlling pore formation in porous silicon is a complicated mix of electronic and chemical factors, and it has been explored and discussed in detail [22]. The parameters of electrolyte composition, dopant type and concentration, applied voltage, temperature, and light intensity all play a role, and many competing mechanisms can be at play in a given experiment. However, there are some general shared features: (i) pores nucleate uniformly and with no particular order on the silicon surface, unless the wafer has been specifically pre-patterned; (ii) current flows preferentially near the pore bottoms; (iii) the pore walls become passivated, leading to dissolution of silicon primarily at the porous silicon/crystalline silicon substrate interface; (iv) once formed, the pores do not redistribute or reconstruct; and (v) all samples contain a distribution of pore diameters rather than a single pore size.

A wide range of pore diameters are accessible in the porous Si electrochemical system [23]. A dramatic example is shown in Figure 1.5, where the pore morphologies generated from n-type (phosphorus-doped) and highly doped p-type (boron-doped) silicon are compared. Both of those samples were generated using identical etching conditions (electrolyte, current density, and etch duration). The morphology of porous silicon derived from n-type wafers tends to consist of macropores, highly doped p+, p++, or n+ samples are mesoporous, and p-type silicon yields meso- to micro-pores. These different morphologies are the result of different pore formation mechanisms. Many mechanisms are thought to contribute to the electrochemical pore growth process in silicon, and the morphology resulting from a given experiment is usually determined by a combination of several of these [23]. Table 1.2 summarizes the types of silicon, the morphologies, and the presumed operative mechanisms for the main types of

a) 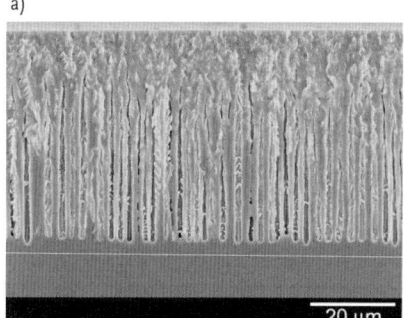 b)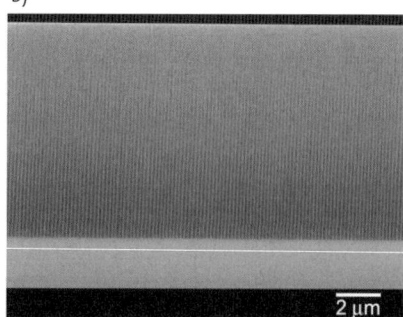

Figure 1.5
Cross-sectional electron micrograph images showing the effect of dopant on pore texture in porous silicon films. Both of these films were prepared under the same conditions of current density ($50\,mA\,cm^{-2}$), etch time (300 s), and electrolyte composition (3:1 aqueous HF:ethanol). Both samples are etched on the (100) face of the crystal. (a) This is an n-type, luminescent sample, etched by back side illumination. The macropores propagate primarily in the <100> direction of the crystal. The porous layer in this image is approximately 50 μm thick. (b) This is a highly doped p^{++}-type sample. Note the pores are so small that they are not well resolved in this image, although the general top-to-bottom texturing reflects the <100> pore propagation direction. This porous layer is 9 μm thick. Preparation of these samples is described in the experiments of Chapter 2. Images provided by Luo Gu, UCSD.

Table 1.2 Pore morpholgies and formation mechanisms depend on dopant.

Pore type	Silicon type[a]	Mechanism	Experiment number[b]
Micropores	p	Crystallographic face selectivity, enhanced electric field, tunneling, quantum confinement	2.1
Mesopores	p+, p++, n+	Enhanced electric field, tunneling	2.2
Macropores	n	Space-charge limited	2.3

a) The superscripts "+" and "++" refer to the level of doping. "+" corresponds to resistivity values of 0.1–0.01 Ω cm, "++" corresponds to 0.01–0.001 Ω cm or less.
b) Refers to Experiment number in Chapter 2. Assumes electrochemical etching in HF-containing alcohol/water electrolytes.

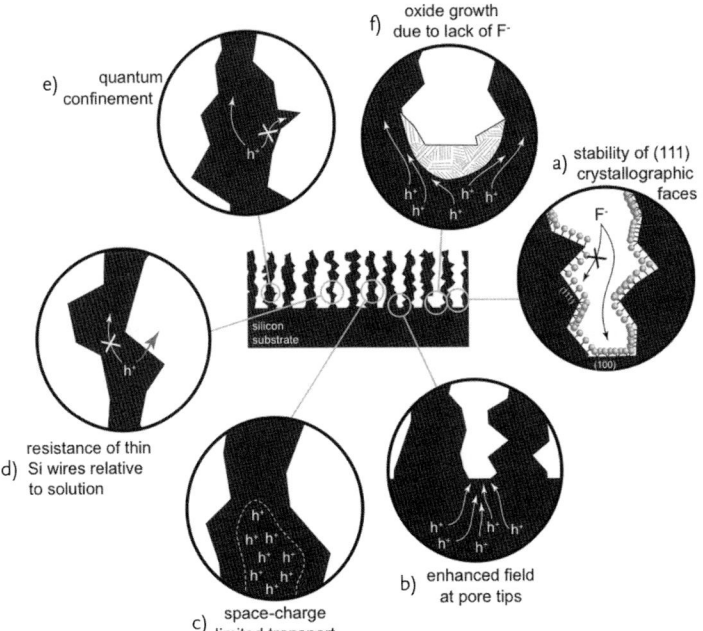

Figure 1.6
Schematic depicting the more important pore-forming mechanisms in porous silicon. (a) The (100) crystallographic face contains strained Si–H bonds, and it tends to be more prone to dissolution compared to other faces. By contrast, the (111) face contains Si–H bonds that are perpendicular to the surface and more stable. The differential reactivity of the crystal faces leads to "crystallographic" pores that propagate primarily in the <100> direction. (b) The high radius of curvature at the tip (i.e., the bottom) of a pore generates a region of enhanced electric field that attracts valence band holes. (c) The space-charge region is a region in which carriers are depleted due to band bending at the silicon/electrolyte interface. It increases with decreasing dopant density, so this mechanism is a primary determinant of macropore size for low-doped n-type Si. (d) As the diameter of a silicon filament decreases, its resistance for transport of valence band holes increases. At a critical filament diameter (typically a few nm for p-type silicon), injection of the hole into the solution becomes more favorable, and holes do not propagate further down the length of the nanowire. This mechanism is responsible for the lack of electrochemical dissolution of a microporous layer, once it is formed. (e) The increased band gap resulting from quantum-confinement excludes valence band holes from these smallest regions of the porous silicon matrix. (f) If there are no fluoride ions available at the silicon/solution interface, silicon oxide forms at the interface. Valence band holes are then excluded from this region and they continue to oxidize the silicon/porous silicon interface. This causes pore widening and, ultimately, lift-off of the porous layer (electropolishing).

etches; Figure 1.6 depicts some of the more important mechanisms. In general they can be split into either physical or chemical phenomena.

1.6.1
Chemical Factors Controlling the Electrochemical Etch

The oxidizing equivalents that start the process are valence band holes, driven to the surface by the applied electric field, and by diffusion. The migration of electrons and holes is influenced by the pore morphology; sharp pore tips generate enhanced electric fields that attract charge carriers. Once a valence band hole makes it to a surface Si atom, the atom is susceptible to attack by nucleophiles in the solution, primarily F^-, and H_2O. A simplified reaction mechanism is given in Figure 1.7

The chemistry that occurs at a silicon surface during electrochemical corrosion involves a competition between Si–O, Si–F, and Si–H bond formation. Si–O bonds are chemically attacked by F^-, and significant quantities of Si–O species only form under conditions in which the diffusion of F^- to the silicon surface cannot keep up with the rate of delivery of valence band holes. Such a condition exists, for example, during electropolishing. Electropolishing involves the complete dissolution of silicon without pore formation, and it is observed when the current density is large, or when the HF concentration in the electrolyte is low. When the concentration of HF in the electrolyte is low, oxidized silicon atoms are generated at the surface too rapidly to be attacked by F^-, and water molecules take over the

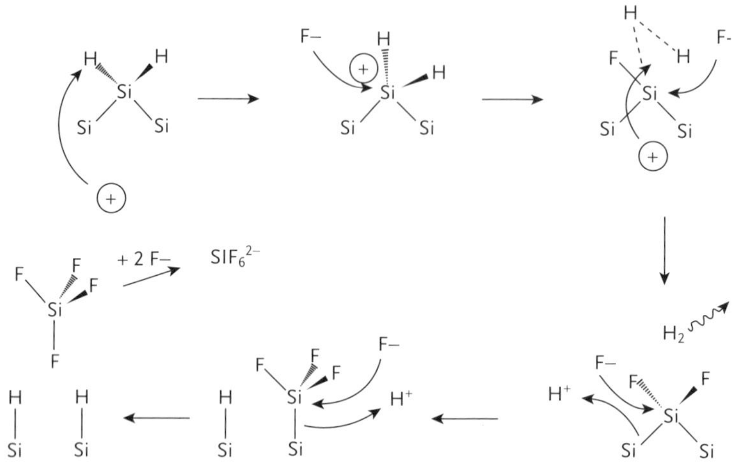

Figure 1.7
Simplified mechanism for electrochemical etching of silicon in fluoride-containing solutions, after [8].

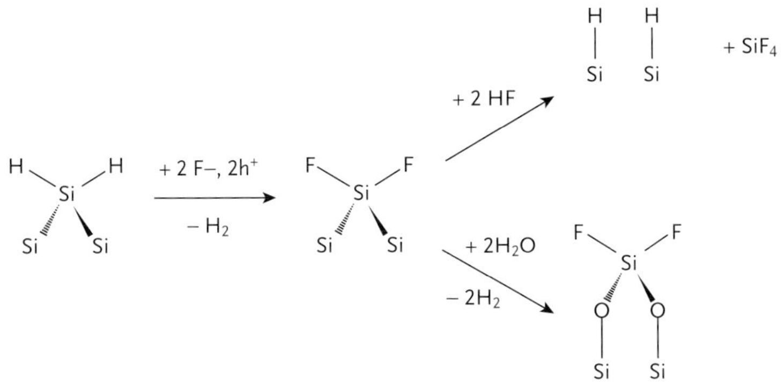

Figure 1.8
Electrochemial oxidation of silicon can take two paths, depending on the availability of fluoride ion in the electrolyte. The upper branch depicts oxidation of silicon when excess HF is present. If the corrosion current exceeds the rate of HF diffusion, water attacks the surface and an insulating oxide is generated [23].

role of nucleophile (Figure 1.8). The reaction mechanism shifts to Si–O formation, and the reaction stoichiometry transitions from 2-electron (Equation 1.9) to 4-electron (Equation 1.6). The lack of fluoride ions means that the oxide cannot be removed from the surface; this insulating oxide terminates propagation of the pore. The valence band holes are then required to move into the porous silicon matrix to oxidize a Si atom that is accessible to the fluoride ions in the electrolyte solution. The result is thinning of the silicon filaments close to the porous silicon/crystalline silicon interface, undercutting the porous silicon layer. An example of the utility of this reaction for generating free-standing, or "lift-off" films of porous silicon is given in Experiment 4.1.

When a silicon wafer is dipped in HF solution, the oxide dissolves and the surface becomes terminated with H atoms. This is puzzling if one considers that Si–F is the most thermodynamically stable bond in all of silicon chemistry; the relative strength of bonds increases in the order Si–H < Si–O < Si–F (Table 1.1). In fact, through the 1980s it was commonly thought that the surface of a silicon wafer cleaned with HF became terminated with Si–F species [24]. It was not until the detailed XPS and FTIR studies of Eli Yablanovitch, Yves Chabal, and others in the 1980s, that the predominance of the hydrides SiH, SiH_2, or SiH_3 was established [25–29]. The reason for the seeming discrepancy is that Si–F bonds are highly polarized due to the large electronegativity of fluorine, and an Si–F surface species is much more susceptible to attack by nucleophiles than an Si–H species. So in a sense, the Si–F bond is too strong; if a fluoride ion attaches to a silicon atom, that atom is rapidly attacked by additional fluoride ions and removed from the surface (Figure 1.7). The surface silicon atoms capped with H atoms thus predominate on the surface during electrochemical etch: hydrogen is much less electronegative than fluorine, and a surface Si–H species is less susceptible to nucleophilic attack. At the end of an HF etch, porous silicon samples contain primarily Si–H, SiH_2, and SiH_3 surface groups and only traces of O or F.

1.6.2
Crystal Face Selectivity

The most obvious morphological characteristic in most porous silicon samples is the tendency for pores to align along the <100> direction of the crystal (see the next section for definitions of these three-number codes, referred to as Miller indices). This is primarily a chemical effect; so-called "crystallographic pores" are a consequence of the stability of the various crystal faces towards chemical attack. For example the hydrogen-terminated (111) face of silicon is the most stable, with hydrogen atoms bonded directly above a silicon atom (Figure 1.9). Much of the internal nanostructure in porous silicon displays small domains resembling (111) facets.

1.6.3
Physical Factors Controlling the Electrochemical Etch

Physical factors that determine pore morphology primarily involve the electronic properties of the semiconductor: the band structure, the type and concentration of the dopants, the influence growing pores exert in shaping the electric field distribution within the wafer, and quantum confinement effects in the small features identified by the etch. The most important factor at play in pore formation is the availability of valence band holes. This is determined primarily by the dopants, but it is also influenced by illumination, HF concentration, and the magnitude of the applied electric

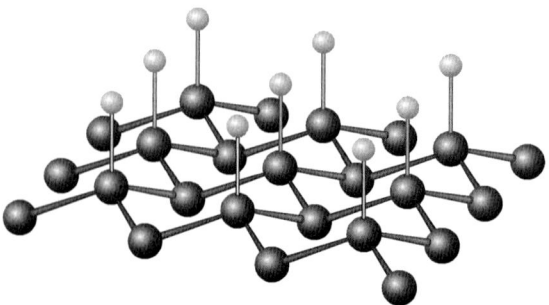

Figure 1.9
The (111) face of a silicon crystal, capped with hydrogen atoms.

field. Whereas crystallographically oriented pores generally form at low current densities and can appear as facet-like structures in cross-sectional scanning electron microscope images, formation of "current-line-oriented-pores", or "current pores" occurs at higher current densities, and these pores tend to be oriented perpendicular to the surface plane of the wafer. The reason for this is that the equipotential surfaces tend to be parallel to the surface of the wafer, yielding a "path of least resistance" for valence band holes in the perpendicular direction. When the "path of least resistance" for a carrier is sideways through the wall of a pore, pore branching occurs. Current pores (either main or branches) are often nucleated at crystallographic pores [30, 31].

In p-type silicon there is a surplus of valence band holes, and the etch is not limited by their availability. In n-type silicon, however, the scarcity of valence band holes limits the number density of pores in a given area of exposed wafer. During an active etch, there exists a zone near the silicon/electrolyte interface where valence band holes are highly depleted, known as the space-charge region. The pore and wall size is determined by the space-charge region in n-type silicon. The electronic properties of silicon also set the rate and conditions needed for the etch, which are discussed later in this chapter.

1.7
Resume of the Properties of Crystalline Silicon

1.7.1
Orientation

Crystalline silicon possesses a diamond lattice structure, with each silicon atom bonded to four other silicon atoms in a tetrahedral coordination environment. The unit cell is shown in Figure 1.10. Common crystal faces are

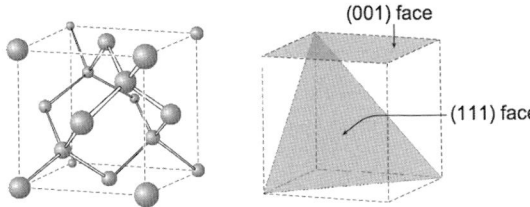

Figure 1.10
The unit cell of silicon. Two of the important crystal faces, with their Miller indices, are shown at the right. The faces of the cube (outlined with the dashed lines) are all equivalent to (100) faces, due to the cubic symmetry of silicon. The cube diagonals represent (111) faces.

defined by three-number codes, known as Miller indices. The numbers in a Miller index correspond to the inverse of the x, y, and z-axis positions in the unit cell that are intersected by the plane of interest. For example, taking the left rear corner of the cube in Figure 1.10 as the origin of a right-hand coordinate system, the crystal face indicated as the (111) face intersects the cube axes at $x = 1$, $y = 1$, and $z = 1$. The Miller indices for this plane are thus $1/x$, $1/y$, $1/z$, or (111). A plane parallel to an axis intersects it at infinity, so the Miller index would be the reciprocal of infinity, or zero. For example the face labeled with Miller indices (001) in Figure 1.10 intersects the x-axis at infinity, the y-axis at infinity, and the z–axis at 1. Since silicon has cubic symmetry, all six faces of the unit cell are equivalent, and the (100), (010), and (001) faces can all be referred to as (100) faces for convenience. Porous silicon is usually prepared from (100) wafers; that is, wafers polished on the (100) crystal face. Pores display a natural tendency to propagate perpendicular to this face, so when porous silicon is etched from (100) polished wafers, the pores drill vertically into such a wafer.

Crystallographic directions are indicated by defining a vector within the unit cell whose components are resolved with respect to the unit cell axes. The coordinates of the vector are placed in square brackets if a particular direction is being defined, or in angle brackets if a crystallographically equivalent family of directions is being defined. For example, the <100> direction in the cubic unit cell of Figure 1.10 corresponds to the equivalent directions defined by vectors x, y, z of 100, 010, 001 as well as the negative directions $\bar{1}00$, $0\bar{1}0$, and $00\bar{1}$. These individual directions would be written [100], [010], [001], [$\bar{1}$00], [0$\bar{1}$0], and [00$\bar{1}$], respectively. Thus we would say that the pores propagate primarily in the <100> direction when a wafer polished on the (100) face is electrochemically etched.

1.7.2
Band Structure

The electronic structure of a molecule can be described as an interaction of atomic orbitals from the individual atoms, which combine to make what chemists call molecular orbitals. Similarly, the electronic structure of a solid derives from atomic orbitals to form what the physicists call energy bands. A key feature of a semiconductor is that the bonding extends over a large distance, often over the entire crystal. The electrons in these orbitals are highly delocalized, making it more convenient to describe bonding in terms of the atomic periodicity in the crystal rather than discrete atomic positions. A complete description is beyond the scope of this book, and we will just give the salient features here. For more detail, there are some good references that describe energy bands from either a chemical [32, 33] or a physical [34] perspective.

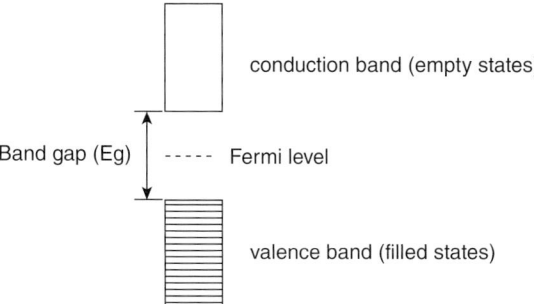

Figure 1.11
Simplified energy band diagram for a semiconductor. The conduction band consists of a series of mostly empty orbitals, while the valence band is made up of orbitals that are mostly filled with electrons. The y-axis in this diagram represents energy, and the band gap represents the energy separating the conduction from the valence band. Electrons move through the empty orbitals of the conduction band and holes move through the electron vacancies in the valence band. The Fermi level represents the average energy of the mobile charge carriers (both conduction band electrons and valence band holes) in the system.

A simplified energy band diagram for a semiconductor is given in Figure 1.11. The band of states (molecular orbitals) that are occupied by electrons is referred to as the valence band, and the band of empty states is the conduction band. The band gap, or energy gap, is the energy separation between the valence band and the conduction band. The valence band derives its name from the fact that it is occupied by what were the valence electrons of the individual atoms making up the solid. The deeper, or core level electrons are tied to the individual atoms and do not contribute significantly to electrical properties such as conduction. So, for the purposes of charge transport, or conductivity, we are only concerned with these charges that are free to move through the solid, and they are called the mobile charge carriers.

1.7.3
Electrons and Holes

Electrons can be placed in the empty orbitals of the conduction band. Since these electrons are free to move throughout the solid, they contribute to conductivity and thus the origin of the name "conduction band." If an electron is removed from the valence band, it leaves behind a positively charged empty site. This vacancy is referred to as a "hole." When a neighboring electron in the valence band moves into this empty site, the space it vacates becomes a new empty site. This effectively moves the hole in the direction opposite to the travel of the electron. Since the net result is a positive charge moving in the opposing direction, it is convenient to refer to this type of conduction as "hole conduction". Thus holes exist and travel in the valence band and electrons travel in the conduction band. Both of these mobile charge carriers can contribute to conductivity in a semiconductor. The Fermi energy is defined as the average energy of all of the mobile charge carriers in the system (both electrons and holes), and its value is represented by the Fermi level, shown with a dashed line in Figure 1.11.

Conductivity is related to the concentration of mobile charge carriers in a semiconductor, and there are four basic ways to increase these concentrations: First, you can heat the sample. This will thermally promote electrons from the valence band to the conduction band, generating an equal number of valence band holes and conduction band electrons. At any temperature greater than absolute zero, there are a finite number of electrons with sufficient energy to jump the band gap. There is always a small contribution to conductivity from this thermal process, and it is the reason that sensitive semiconductor-based light detectors, such as photodiodes and CCDs (charge coupled devices) need to be cooled to reduce noise in their signals. Second, you can shine light on the sample. Light with energy greater than the bandgap energy will promote an electron from the valence band to the conduction band, simultaneously generating a conduction band electron and a valence band hole. Third, you can dope the sample with an impurity that donates one of its electrons to the conduction band of the semiconductor or accepts an electron from the valence band. Fourth, you can inject electrons or holes from a material placed in contact with the semiconductor. For example, a transistor operates by injection of carriers from a "gate" electrode into a region of semiconductor that is devoid of carriers. The next sections discuss the two mechanisms for increasing carrier concentration that are most relevant to this book: photoexcitation and doping.

1.7.4
Photoexcitation of Semiconductors

The energy to promote an electron from the valence to the conduction band can also be supplied by a photon of light, provided its energy exceeds the band gap energy. There are well-defined selection rules for this process, and one of the key requirements in crystalline silicon is that the photoexcitation event be accompanied by a lattice vibrational quantum, that is, a phonon. Semiconductors that require phonon absorption for photoexcitation are *indirect gap* semiconductors, and those that do not are *direct gap* semiconductors. Silicon is an indirect gap semiconductor. This phonon constraint has important ramifications in processes that involve absorption of light, such as photoetching of n-type silicon or direct photopatterning of porous silicon during fabrication. Because an indirect gap transition is a two-body problem (photon and phonon transitions must take place simultaneously), it has a low probability. This translates into a small extinction coefficient. For example, on average a red photon will penetrate about 7 µm into silicon before it is absorbed, whereas, in gallium arsenide, a direct gap semiconductor, the penetration depth is less than 700 nm.

1.7.5
Dopants

Dopants, sometimes called impurities, are elements added to a semiconductor to increase its conductivity. The element chosen to act as a dopant generally has one extra or one less valence electron than the semiconductor. For example, phosphorus lies to the right of silicon in the periodic table, and so it has one more valence electron than silicon. It is added to the molten Si during production, and it replaces a Si atom in the crystal lattice. In the language of solid state chemistry, this is referred to as a substitutional defect. The extra electron is donated to the conduction band, increasing the conductivity of the semiconductor. Similarly, a boron atom substituted into the lattice increases conductivity by donating a hole to the valence band.

Electrons and holes are referred to as charge carriers, because they carry charge (current) through the semiconductor. Keep in mind the term "holes" here is the electronic term, referring to an electron vacancy in the valence band rather than a physical void. Dopants generate an excess of one type of charge carrier; the "n" in "n-type" refers to a negatively charged carrier, that is, an electron, while the "p" in "p-type" refers to the positively charged hole. Undoped silicon has an equal number of electrons and holes due to intrinsic ionization of the pure material, and it is called "intrinsic" silicon. Dopants for Si are generally either phosphorus (for n-type doping) or boron (for p-type doping). Due to the solubility limits of phosphorus, highly doped n-type wafers sometimes use antimony (Sb) as a dopant. Do not confuse the element symbol for phosphorus (P) with the indication of dopant type; phosphorus-doped silicon is n-type. Manufacturers will usually spell out the element name to avoid confusion. For example, a packing label on a wafer box will say "type: P, dopant: boron" to indicate a p-type lot, or "type: N, dopant: phosphorus" to indicate n-type wafers. You should check for both descriptors to be sure you are using the intended material.

Under normal conditions, electrons and holes are in equilibrium with each other in a semiconductor, and you cannot increase the concentration of one of these carrier types without decreasing the concentration of the other. If you increase the concentration of electrons (by adding a phosphorus dopant, for instance), some of the excess electrons will recombine with the available holes to establish a new equilibrium. There is a form of the law of mass action, equivalent to the equilibrium relationship between H^+ and OH^- in aqueous solutions, described by Equation 1.10:

$$n \cdot p = n_i^2 \qquad (1.10)$$

where n and p are the concentrations of mobile electrons and holes, respectively, and n_i is a constant, representing the equilibrium constant for electrons and holes in the semiconductor. It is temperature dependent; for silicon at room temperature, $n_i = 1.45 \times 10^{10}\,\mathrm{cm}^{-3}$. This value represents the equilibrium concentration of electrons or holes in a pure (undoped, or intrinsic) silicon crystal.

1.7.6
Conductivity

Conductivity measures the ability of a material to carry electrical current, which is the transport of charge. Resistivity is the inverse of conductivity, measuring the propensity of the material to impede charge transport. For any material conductivity generally relies on two parameters: the concentration of the charge carriers (electrons and holes), and the mobility of these carriers. Whereas any material has plenty of electrons in the valence and core levels of its atomic constituents, if the electrons are not free to move from one spot to the next the material cannot conduct electricity. The basic relationship between conductivity (σ), carrier mobility (μ_n, μ_p), and carrier concentrations (n, p) is given by Equation 1.11

$$\sigma = q(\mu_n n + \mu_p p) \qquad (1.11)$$

Representative room-temperature mobility values for silicon are $\mu_n = 1300\,\mathrm{cm}^2\,\mathrm{V}^{-1}\,\mathrm{s}^{-1}$ and $\mu_p = 450\,\mathrm{cm}^2\,\mathrm{V}^{-1}\,\mathrm{s}^{-1}$. An example of the use of Equations 1.10 and 1.11 to calculate carrier concentrations in a sample of known resistivity is given in Experiment 1.1.

1.7.7
Evolution of Energy Bands upon Immersion in an Electrolyte

When a metal contact is made to a silicon wafer, a small number of electrons will pass between the two materials to equilibrate the work functions. If the work functions of the two materials are of the appropriate values, a build-up of fixed charge at the interface results. This interfacial charge acts as a barrier for transport of additional electrons. Differences in dielectric constant and density of states between the two materials lead to an asymmetric barrier, resulting in the unidirectional, or rectifying electron transport behavior commonly observed with a diode. For a metal/semiconductor contact, the built-in field is referred to as a Schottky barrier.

A similar situation to Schottky barrier formation occurs when a silicon electrode is immersed in an electrolyte. Charge equilibration between the two phases leads to a barrier that either blocks or allows current to flow, depending on the direction of the current. The energy band diagrams of Figure 1.12 show the situation before and after immersion of n-type and p-type silicon electrodes into electrolyte. The barrier is depicted as a bending of the conduction and valence bands in the vicinity of the interface. In the energy band diagrams, it is energetically favorable for holes to move upwards along the band lines because of their positive charge, and electrons

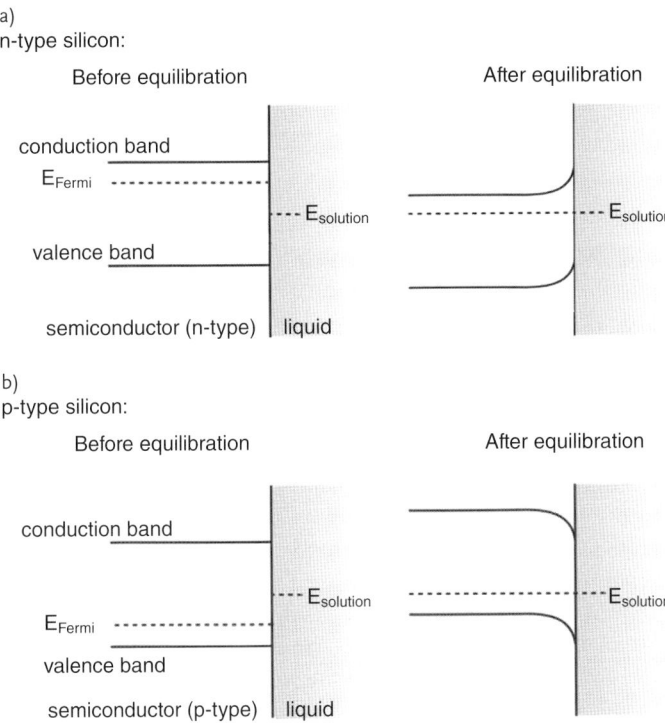

Figure 1.12
(a) Equilibration of charge at the semiconductor/electrolyte interface for an n-type semiconductor. The evolution of the energy band structure that occurs upon immersion in a liquid is depicted. The y-axis in this diagram is energy, and the spatial dimension through the semiconductor/electrolyte interface is depicted along x. The energy bands bend in the vicinity of the interface as a result of charge equilibration between the semiconductor and the liquid phase.
(b) Equilibration of charge at the semiconductor/electrolyte interface for a p-type semiconductor. The energy bands bend as a result of charge equilibration between the semiconductor and the liquid phase.

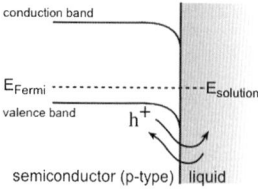

Figure 1.13 Charge transport processes at the semiconductor/electrolyte interface for a p-type semiconductor. The increase in hole current across the interface under forward bias (silicon positive relative to the Pt counter-electrode in solution) is depicted. The bent energy bands provide an impediment to charge transfer that is overcome with a small applied bias.

will spontaneously move downwards. A common mnemonic is to think of holes as bubbles that float upwards and electrons as balls that roll downwards along the energy band lines in these diagrams.

1.7.8
Charge Transport at p-Type Si Liquid Junctions

Valence band holes are the important charge carriers that lead to electrode corrosion. For a p-type silicon sample, these are the majority carriers. With no applied bias, the rate of hole transport in either direction across the interface is small, as depicted in Figure 1.13. When a positive bias is applied to the silicon electrode, the barrier to interfacial charge transport is reduced and valence band holes begin to accumulate at the interface. In the language of solid state electronics, we would say that the diode is in forward bias. The holes participate in the corrosion reactions of Equation 1.6 or 1.9, and porous silicon begins to form.

If the potential of the silicon electrode is adjusted to be negative of the solution potential, the diode goes into reverse bias. The situation depicted in Figure 1.14 occurs. The barrier to hole transport to the interface increases, and the corrosion reaction is effectively shut off.

1.7.9
Idealized Current–Voltage Curve at p-Type Liquid Junctions

The idealized current–voltage curve for a p-type silicon wafer in an HF electrolyte is shown in Figure 1.15. The p-silicon/electrolyte junction is in forward bias during etch, so the current will increase exponentially with even a small increase in applied voltage. The mathematical relationship for the current–voltage curve of an ideal diode is also given in the figure. Note

1.7 Resume of the Properties of Crystalline Silicon

Figure 1.14
Blocking of charge transport across the semiconductor/electrolyte interface under reverse bias conditions for a p-type semiconductor. A decrease in the interfacial transport of holes occurs when the semiconductor is placed under reverse bias conditions. In this situation, only a small leakage current (reverse saturation current) flows into solution.

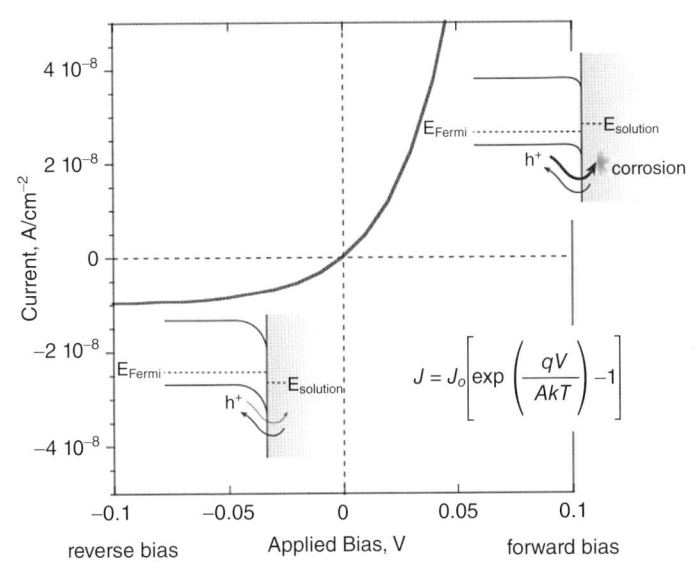

$$J = J_0\left[\exp\left(\frac{qV}{AkT}\right) - 1\right]$$

Figure 1.15
Ideal Diode Law applied to electrocorrosion of a p-type liquid junction. Under forward bias (right-hand side of the plot), the corrosion (hole) current turns on exponentially. Under reverse bias conditions, the hole current across the semiconductor/electrolyte interface is limited to the value of the reverse saturation current density, J_0. Etching proceeds when the system is in forward bias.

the similarity of this curve to the first part of the curve in Figure 1.3 (in the porous silicon formation region). The ideal curve of Figure 1.15 is only approximately correct for electrochemical corrosion of silicon. As porous silicon forms, the defect density at the interface increases, degrading the quality of the junction.

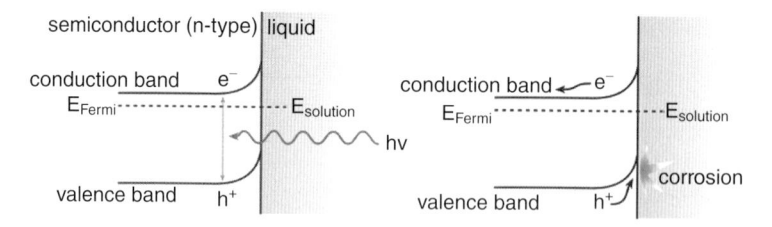

Figure 1.16

n-type Si must be illuminated in order to drive electrocorrosion. Light generates electron–hole pairs. For n-type Si, the built-in field sweeps the valence band holes to the semiconductor/electrolyte interface, where they supply the oxidizing equivalents for the corrosion reaction.

1.7.10
Energetics at n-Type Si Liquid Junctions

Holes are the important charge carriers that lead to electrode corrosion. Because the band bending at an n-type semiconductor electrode is opposite to that of a p-type electrode (Figure 1.12), hole current at the n-type interface is blocked by the junction. The majority of carriers in an n-type semiconductor are electrons; to generate a sufficient number of holes to carry out the corrosion reaction, the semiconductor must be illuminated. Light generates electron–hole pairs near the semiconductor interface, and the built-in field sweeps the holes to the surface. This is depicted schematically in Figure 1.16.

1.7.11
Idealized Current–Voltage Curve at n-type Liquid Junctions

The idealized j–V curve for an n-type semiconductor in contact with an electrolyte is shown in Figure 1.17. Two traces are shown, one for an electrode in the dark and the other for one under illumination.

1.8
Choosing, Characterizing, and Preparing a Silicon Wafer

When purchasing crystalline silicon wafers (Figure 1.18), you must specify three main properties: the cut (crystal orientation) of the wafer, dopant type, and resistivity. All three of these parameters are crucial in determining the size, number, and direction of the pores that will be generated in your porous silicon samples. In addition, there are several other properties that

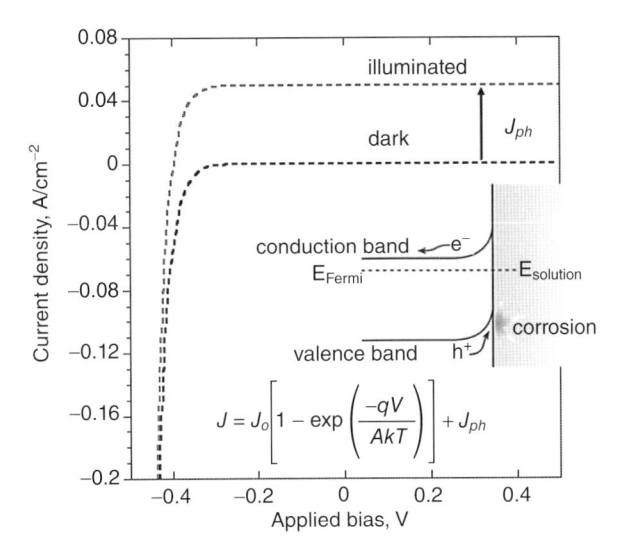

Figure 1.17
Current–voltage curves for an n-type liquid junction. For an n-type junction, the corrosion (hole) current is blocked by the diode. Corrosion is induced by illumination of the semiconductor, which generates an excess of holes. The built-in field of the diode draws the holes to the interface, where they contribute to corrosion of the electrode surface. The equation for the Ideal Diode Law, incorporating the effect of photocurrent (J_{ph}), is shown. Under reverse bias conditions, the hole current across the semiconductor/electrolyte interface is limited to the sum of the reverse saturation current and the photocurrent, $J_0 + J_{ph}$.

you should understand when working with silicon. Those parameters usually given by the wafer vendors are presented in Table 1.3. We described the key structural and electronic properties underlying the parameters relevant to the preparation of the various porous silicon morphologies in the previous sections. In the following we describe the techniques used to characterize the crystalline silicon starting material. There are many good references that provide more detailed discussion. One of the most useful is [35].

1.8.1
Measurement of Wafer Resistivity

One of the biggest problems in the measurement of resistivity of a material is contact resistance between the probe leads and the sample. For example, a silicon wafer usually has about 7 nm of native oxide on its surface. If you place the two leads of an ohmmeter (multimeter set to "Ohms" setting) on the surface of a silicon wafer, the resistance value you measure is

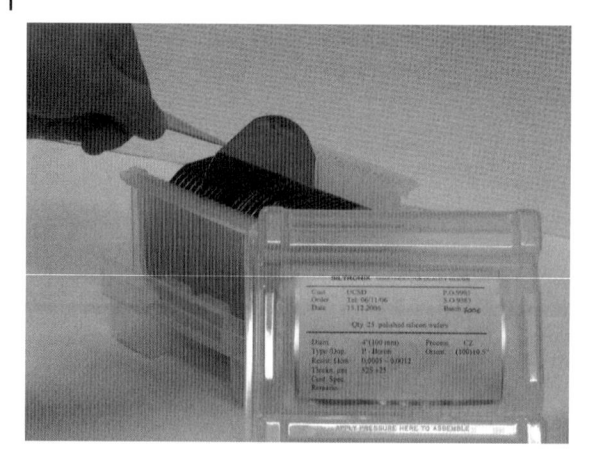

Figure 1.18

Crystalline silicon wafers, fresh out of the box from the vendor. The three most important properties are crystal orientation, dopant type, and resistivity. Wafers are generally shipped in polypropylene containers and should be handled with nylon or other plastic tweezers. Impurities from fingerprints or metallic tweezers can influence the outcome of various processing steps such as etching or thermal treatment. These wafers are 6 in diameter, p-type wafers, polished on the (100) surface, doped to a very high concentration with boron. With a nominal resistivity of 0.001 Ω cm, these wafers would be considered degenerately doped, designated p^{++} type Si.

dominated by this highly resistive SiO_2 layer, even if the underlying silicon is highly conductive. The four-point probe method eliminates the effect of contact resistance. The four-point experiment is pictured schematically in Figure 1.19 and photographically in Figure 1.20.

A constant current is passed between the outer two probe tips, while the voltage drop is simultaneously measured between the inner two tips using a high impedance voltmeter. Resistivity is determined from the measured values of voltage drop (V) and current (I)

The sheet resistance (R_s) is calculated from the ratio of current to voltage, multiplied by a correction factor that is related to the sample geometry:

$$R_s = \frac{V}{I} \cdot CF \tag{1.12}$$

Where V is in volts and I is in amps. The value of CF is 4.54 for an infinite sheet (this approximation is valid if the sample is over a factor of 10 larger in both the x- and y- directions than the probe tip spacing). The units of sheet resistance are ohms per square.

Sheet resistance and resistivity (ρ) are related by:

$$\rho = R_s W \tag{1.13}$$

1.8 Choosing, Characterizing, and Preparing a Silicon Wafer

Table 1.3 Typical crystalline silicon wafer parameters specified by the manufacturer.

Parameter	Description	Example
Orientation	Miller index of the wafer, indicates which crystal face is located on the face of the wafer	<100>, polished on the (100) face. Most commonly used for porous Si preparation. Note the direction of pore propagation is primarily in the <100> direction; etching a (100) wafer will yield pores perpendicular to the surface. If you choose a (110) wafer by mistake, you will get a cross-hatching of pores, propagating at angles of 45° from the surface normal.
Type/dopant	Type of doping (n or p) and chemical element used to dope the wafer	P-Boron. Boron is the dopant; it generates a p-type impurity
Resistivity	The electrical resistivity of the wafer, related to the dopant concentration	0.0005–0.0012 Ω cm. Manufacturing is never perfect; a range of values is usually specified. You should always verify with a 4-point probe measurement.
Thickness	Thickness of the wafer	525 ± 25 μm.
Process	Method used to grow the silicon crystal	CZ. Grown using the standard Chockraszki method. Very high quality (and expensive) wafers are grown by the float zone method, designated FZ.

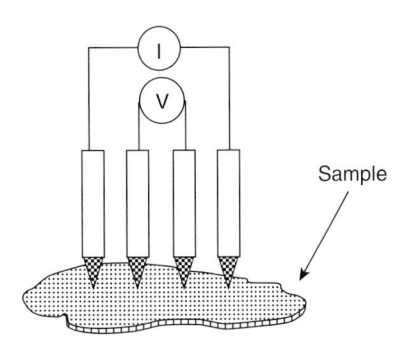

Figure 1.19
Schematic diagram of the four-point resistivity experiment.

Where W is the thickness of the wafer being measured. The usual units of resistivity are ohm cm (Ω cm) so the thickness should be measured in cm. You should not trust the nominal wafer thickness quoted by the manufacturer on the box because thickness can differ significantly from wafer to wafer in a given lot.

1 Fundamentals of Porous Silicon Preparation

EXPERIMENT 1.1: Measure the resistivity of a silicon wafer

In this exercise, you will measure the resistivity of a silicon wafer using a four-point resistivity probe. You will then calculate the dopant concentration in the wafer using the ideal mobility relationship.

Equipment needed:

4-point resistivity probe	Complete probe stations are available from many manufacturers. A simple test station like the Alessi CPS resistivity test fixture with a C4S 4-point probe head (available from Cascade Microtech, www.cascademicrotech.com) is a good option.
Constant current source, voltmeter, ammeter	All-in-one packages, such as the Keithley 2001 multimeter, (www.keithley.com) that source a current, measure the voltage drop, and output a resistance value ("4-wire resistance measurement") are a simple solution. Alternatively, a power supply that can supply a constant current (B + K Precision 1623A, www.bkprecision.com) and two digital multimeters (Fluke 187, www.fluke.com), one configured to measure current, the other to measure voltage, can be used.
Digital micrometer	Mitutoyo 293-832, (www.mitutoyo.com)

Procedure:

1) Measure the resistivity of the sample using the 4-point probe. Make 5 measurements in 5 separate spots on the wafer, and report the average value. Be sure that all of the test spots are at least 10 times as far from the edge of the wafer as the spacing of the probe tips. An example of the connections appropriate for the Keithley model 2001 DMM is shown in Figure 1.20

2) Depending on your specific measurement apparatus, you will have either pairs of voltage and current readings or you will have individual "4-wire resistance" values, where the ratio of V/i has already been calculated for you. Examples of both are shown in the table below.

3) Measure thickness of sample with the micrometer. Note this will scratch the surface of your sample, which can affect the quality of the porous Si formed in the vicinity.

1.8 Choosing, Characterizing, and Preparing a Silicon Wafer

To calculate resistivity:

a) Convert resistance measurement (V/i) to sheet resistance, R_s:

$$R_s = \frac{V}{i} \cdot CF$$

b) where $CF = 4.532$ (infinite sheet approximation)
c) Convert R_s to resistivity:

$$\rho = R_s \cdot W$$

d) where $W =$ your wafer thickness, in cm (measure with micrometer)

Results:

Values are given below for a representative sample.
Sample thickness = 0.0537 cm

	Voltage, V (mV)	Current, i (mA)	V/i (Ω)	Rs (Ω per square)	Resistivity (Ω cm)
1	0.038	8.530	0.00445	0.02019	0.00108
2	0.038	8.529	0.00446	0.02019	0.00108
3	0.038	8.525	0.00446	0.02020	0.00108
4	0.038	8.532	0.00445	0.02019	0.00108
5	0.037	8.545	0.00433	0.01963	0.00105
Avg	0.0378	8.532	0.00443	0.02008	0.0011

Sample resistivity = 1.1 mΩ cm

The simplified relationships for conductivity and carrier concentration are given in Equations 1.10 and 1.11. Using the mobility values $\mu_n = 1300\,\text{cm}^2\,\text{V}^{-1}\,\text{s}^{-1}$ and $\mu_p = 450\,\text{cm}^2\,\text{V}^{-1}\,\text{s}^{-1}$, the intrinsic silicon constant $n_i = 1.45 \times 10^{10}\,\text{cm}^{-3}$, and $q = 1.602 \times 10^{-19}\,\text{C}$, the concentrations of electrons and holes in this sample can be calculated:

$$\sigma = \frac{1}{\rho} = q(\mu_p p + \mu_n n)$$

assuming $n \ll p$,

$$p = \frac{1}{\rho q \mu_p} = 1.3 \times 10^{19}\,\text{cm}^{-3}$$

$$n = \frac{n_i^2}{p} = 17$$

Note, in this very highly doped semiconductor the concentration of free electrons is negligible (~17) and all the current is carried by valence band holes. This calculation assumed $p \gg n$; as you can see, a fair assumption for this particular sample. If the carrier concentrations are closer to each other (low-doped n- or p-type material), you have to solve the problem using the binomial expansion.

If we assume that all the boron dopant atoms in the semiconductor can be ionized to donate a hole to the valence band, the dopant density (in atoms cm^{-3}) is equal to the value of p, or 1.3×10^{19} cm^{-3}. This corresponds to a boron concentration of 200 ppm (200 B atoms per 10^6 Si atoms) in the Si crystal. So, even with very highly doped silicon, the dopant atoms represent a very small fraction of the overall composition.

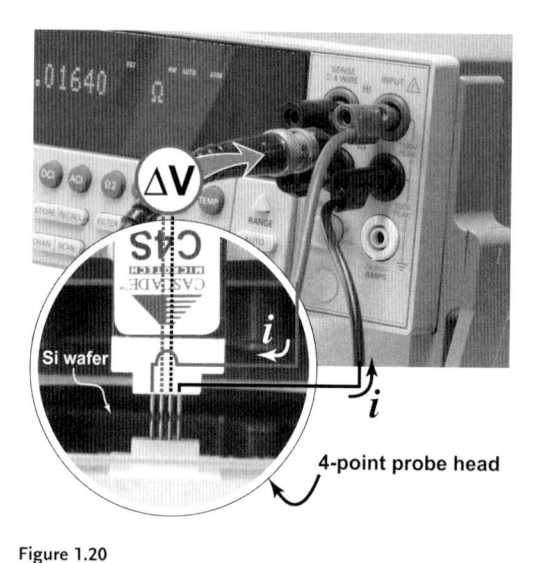

Figure 1.20

Connections for a 4-point probe measurement of wafer resistivity using the Keithley 2001 DMM (www.keithley.com) in the four-wire resistance mode. Inset shows a 4-point probe head in contact with a polished silicon wafer, and the electrical connections are depicted. Specific procedure to measure resistivity of a wafer using this unit: (a) Power on, (b) press "Ω4" button, (c) press "config" button, (d) Speed: "HiAccuracy", (e) filter: 25 (this is the number of measurements to average), (f) resln: "7.5 d", (g) OffsetComp: "on", (h) Exit, (i) put copper foil on the stage, place probe tips in contact, (j) press "rel" to zero reading, (k) replace the copper foil with your sample. Measure your sample in 5 spots. Toggle "filter" button to turn on and off averaging between measurements.

1.8.2
Cleaving a Silicon Wafer

During manufacture, silicon wafers are cut from a large single crystal (known as a boule) with a diamond saw. The angle of the cut is chosen to coincide with a particular crystal face, and this face is then polished to a

1.8 Choosing, Characterizing, and Preparing a Silicon Wafer

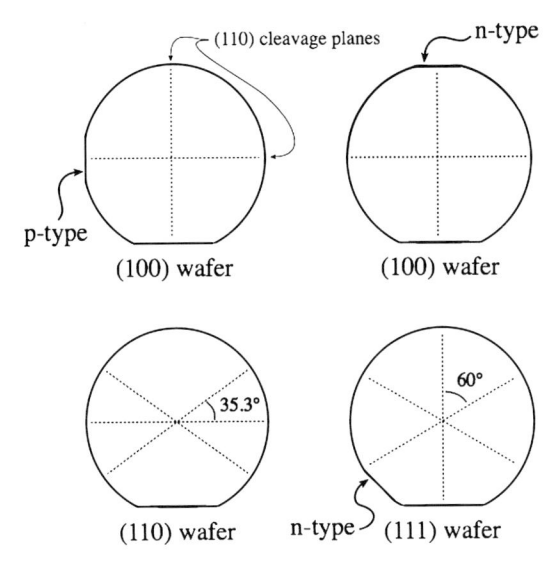

Figure 1.21
Silicon wafers and the meaning of the fiducial flats, as adopted by the SEMI Standards organization (www.semi.org). Dashed lines indicate the location of the (110) natural cleavage planes in the wafer. For n-type, (100) wafers, there are 2 flats: the major flat is parallel to the (110) cleavage plane, the second flat is parallel to the major flat, on the opposite side of the wafer. For p-type, (100), there are 2 flats: the major flat is parallel to the (110) cleavage plane, the second flat is at 90° from the major flat. For n-type, (111), there are 2 flats: the major flat is parallel to (110), the second flat is 45° from the major flat. For p-type, (111), there is only 1 major flat. For wafers with diameters >200 mm, the orientation is indicated with a notch instead of a flat. Adapted from [35].

mirror finish. The wafer is a round disc, though the orientation of other crystal planes relative to the polished face is indicated by flats cut into the wafer, as indicated in Figure 1.21. These markings allow one to find the (110) natural cleavage planes of silicon.

Silicon will break easily along these natural cleavage planes of the crystal. To break a wafer into squares of the appropriate size for the experiments in this book, it helps to scribe a small notch or groove on the back (unpolished) side of the wafer first. A diamond or tungsten carbide scribe can be used (Fisher Scientific, www.fishersci.com, cat # 08-675). The following technique can be used:

1) Place the wafer polished side down on a clean, dust-free tissue paper to protect the surface

2) Place a straight edge on the back side of the wafer, parallel to one of the cleavage planes. The line should be placed to produce a strip of silicon approx 2 cm wide

3) Using the scribe and the straight edge as a guide, score a line. Go over the line several times with the scribe to ensure the groove is deep enough to yield a good cleave

4) Flip the wafer over onto a stack of 10–20 paper towels, place a clean tissue onto the polished face

5) Place a round stick, such as a pencil, on the wafer parallel to and directly over the scored line.

6) Apply gentle, even pressure to the stick until the wafer cracks. It should crack along the length of the score.

7) Repeat the procedure to make 2 cm squares. The squares should be a few mm larger than the O-ring used in the etching cell, to ensure a leak-free seal.

Wafers can also be cleaved without placing a score on the back side, although it takes a little more practice. Place the wafer on top of a single tissue on a hard, flat surface. Using a razor blade, apply firm pressure to the edge of the wafer and pull the blade off the edge until it snaps to the surface of the table. A crack should propagate along the (110) cleavage plane. With a little practice, this is much faster than the scoring method.

1.8.3
Determination of Carrier Type by the Hot-Probe Method

Although the manufacturer will specify the carrier type (n or p), it is sometimes necessary to verify this. The hot-probe measurement is a simple method to distinguish between n- and p-type semiconductors (Figure 1.22).

1.8.4
Ohmic Contacts

In order to etch a silicon wafer, a low-resistance contact must be made to the back side of the wafer. For highly doped substrates ($<0.1\,\Omega\,cm$), a metal foil (aluminum or platinum) placed in contact with the wafer is sufficient. When etching p-or n-type silicon with resistivity ($>0.1\,\Omega\,cm$), the metal forms a rectifying junction, generating a barrier to electron transport that leads to irreproducible and non-uniform porous layers. The rectifying junction acts like a Schottky diode, illustrated in Figure 1.23, with the relevant equivalent circuits shown.

In addition to the rectifying nature of some metal contacts, a metal (e.g., alligator clip or aluminum foil) pressed against silicon will generally have some contact resistance associated with the oxides coating the surfaces of

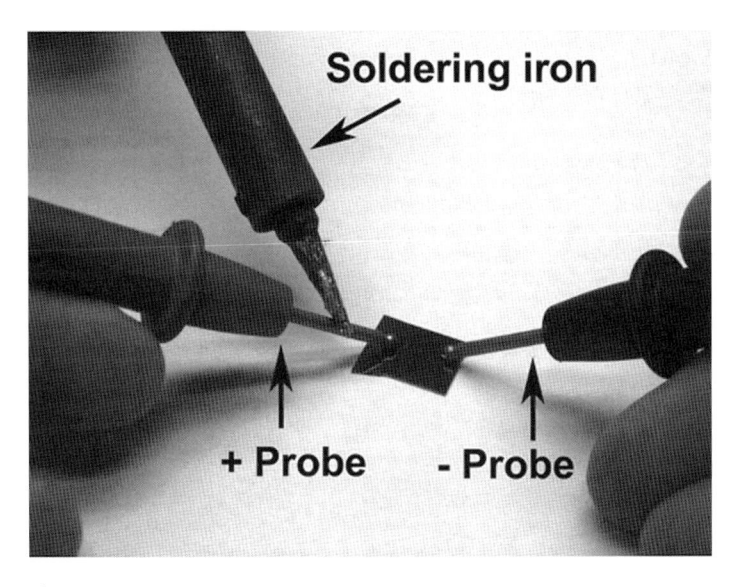

1.8 Choosing, Characterizing, and Preparing a Silicon Wafer | 37

Figure 1.22
Hot probe measurement of carrier type. The sign of the voltage difference between a probe heated with a soldering iron and one at ambient temperature identifies whether a chip is n- or p-type. When the black lead shown on the right is connected to the ground input on the DMM, a positive voltage indicates an n-type wafer; negative voltage indicates p-type. It helps if you have three hands available for this measurement.

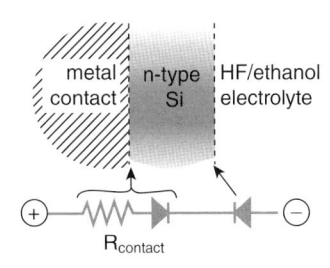

Figure 1.23
The contacting interfaces and the corresponding equivalent circuits for etching of n- and p-type Si. In either case, the diodes corresponding to the metal–Si and the Si–electrolyte junctions are opposing. In order to pass significant current through the p-type semiconductor, the metal–silicon contact must be close to ohmic. In order to pass significant current through the n-type semiconductor, the silicon–electrolyte junction must be illuminated in order to generate photocurrent.

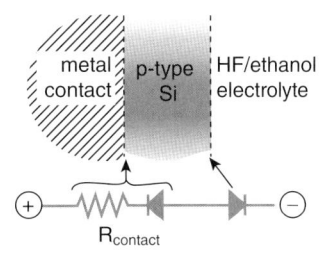

> **EXPERIMENT 1.2: Hot probe determination of the carrier type in a silicon wafer**
>
> In this exercise, you will determine the carrier type in a silicon wafer using a soldering iron and a multimeter.
>
> Equipment needed:
>
> | | |
> |---|---|
> | Digital multimeter (DMM) | For this experiment you will need a meter with a floating ground, to avoid establishing a ground loop with the soldering iron. Use of a battery-powered DMM will do the trick. The Fluke 187 (www.fluke.com) is a good choice. |
> | Soldering iron | Weller SPG40 |
>
> **Procedure:**
>
> Choose either an n- or a p-type piece of Si. Set the DMM to the lowest DC voltage setting (mV on the Fluke 187). Place the two probes on the chip and touch the hot soldering iron to the probe leading to the + input (red V) of the DMM. After a few seconds, note the sign of the voltage reading.
>
> **Results:**
>
> For sample resistivities in the $1\,\Omega\,cm$ range, you should observe a voltage difference of 50–100 mV, depending on how hot the probe gets. For an n-type sample, the sign of the voltage will be positive; for a p-type sample, the sign will be negative. You should move the iron from the + to the − lead of the DMM and observe a change in the sign of the voltage to test the consistency of the measurement – dirty probe tips can give a false reading.

both the metal and the silicon electrode. The practical consequence of either junction formation or contact resistance is that the etching power supply must work harder to push the desired current through the silicon/electrolyte interface. This means it must supply more voltage. The power supplies listed in Table 2.1 can all supply the voltage sufficient for most experiments, though there may be undesired consequences if the contact resistance gets too high. As mentioned above, irreproducible or nonuniform pore morphologies are the most common result. For some of the experiments performed in this book we ignore the effects of the junction and just let the power supply do the work. However, in many cases it is appropriate to make an ohmic contact to the wafer before etching, in particular for p-type silicon.

An ohmic contact is defined as a metal–semiconductor junction that has negligible contact resistance relative to the resistance of the bulk semiconductor. To make such a contact, the interfacial resistance and/or rectifying nature of the silicon/metal interface must be destroyed. It is generally accomplished by evaporating a metal with a Fermi energy appropriate to produce a low Shottky barrier. Alternatively, physical abrasion generates surface defects that act as efficient carrier recombination centers, negating the rectifying properties of the junction.

In order to make a good back-contact to the silicon wafer in an etching cell, you should:

1) Remove the native oxide from the back of the wafer by soaking or rinsing the back side of the wafer in 3:1 48% aqueous HF:ethanol solution for 10–30 s before mounting it in the cell. Rinse with ethanol and dry.

2) Make sure to use a clean (platinum or aluminum foil) back-contact.

3) Make sure the clips used to connect the back-contact and the platinum counter-electrode are clean and in good shape. After time, exposure of copper or brass alligator clips to the corrosive HF fumes will destroy them.

It is a good idea to monitor the voltage being applied by the power supply during etch. It should be between 0.5 and 5 V, typically. If the applied voltage is >5 V – or worse: pushing close to the maximum voltage the source can supply – you will probably get irreproducible results (lower porosity than expected, thinner film, unusual layered structures, different optical properties, etc.) If you see strange results, you can troubleshoot the etching rig by systematically testing all the contacts using a digital multimeter. If it is clear that the problem is the metal–silicon contact, you need to make an ohmic contact to the wafer.

1.8.4.1 Making an Ohmic Contact by Metal Evaporation

For an ohmic contact to n-type silicon, a gold–antimony (Au–Sb) alloy containing 0.1% Sb is evaporated onto the wafer (remove the native oxide from the wafer with HF solution first, see above), and then the wafer is annealed at 300 °C in an inert (vacuum, N_2 or Ar) atmosphere for 3 h.

For an ohmic contact to p-type silicon, a thin layer of aluminum is placed on the wafer by either sputter-coating or by e-beam evaporation. A standard recipe for sputter-coating is:

1) Clean the side to be contacted using an RCA etch.

2) Sputter-coat the wafer using a commercial sputter-coater (Rotation: 65, pressure: 2.6–2.7 mtorr, Ar gas flow: 35 SCCM, power: 150 W, time: 15 min (yields approx. 200 nm thick film).

A standard recipe for electron beam evaporation:

1) Clean the side to be contacted using an RCA etch.
2) Deposit at a rate of $0.2\,\text{nm}\,\text{s}^{-1}$ to 200 nm thickness.

For either e-beam evaporation or sputter coating, samples must be thermally annealed in an inert atmosphere or in vacuum (150 °C for 2 min, then 450 °C for 15 min).

1.8.4.2 Making an Ohmic Contact by Mechanical Abrasion

A quick, brute force method of making an ohmic contact to either n- or p-type Si is to place a small amount of 1 : 1 gallium/indium (Ga/In) eutectic on the back side of the wafer and work it into the surface by scratching with a razor blade or carbide scribe until the entire back surface is shiny. The low work function of the eutectic makes it a good ohmic contact to n-type Si, although scratching generates so many interfacial defects that it works pretty well for p-type too. Gallium and indium are toxic, so the samples must be handled with care. After application, the eutectic can be coated with silver paint (SPI Supplies, # 05001-AB, www.2spi.com) to protect it.

References

1 Uhlir, A. (1956) Electrolytic shaping of germanium and silicon. *Bell Syst. Tech. J.*, **35**, 333–347.
2 Gupta, P., Colvin, V.L., and George, S.M. (1988) Hydrogen desorption kinetics from monohydride and dihydride species on Si surfaces. *Phys. Rev. B*, **37** (14), 8234–8243.
3 Gupta, P., Dillon, A.C., Bracker, A.S., and George, S.M. (1991) FTIR studies of H_2O and D_2O decomposition on porous silicon. *Surf. Sci.*, **245**, 360–372.
4 Dillon, A.C., Gupta, P., Robinson, M.B., Bracker, A.S., and George, S.M. (1990) FTIR studies of water and ammonia decomposition on silicon surfaces. *J. Electron. Spectrosc. Relat. Phenom.*, **54/55**, 1085–1095.
5 Dillon, A.C., Robinson, M.B., Han, M.Y., and George, S.M. (1992) Diethylsilane decomposition on silicon surfaces studied using transmission FTIR spectroscopy. *J. Electrochem. Soc.*, **139** (2), 537–543.
6 Anderson, R.C., Muller, R.S., and Tobias, C.W. (1990) Investigations of porous Si for vapor sensing. *Sens. Actuators*, **A21-A23**, 835–839.
7 Canham, L.T. (1990) Silicon quantum wire array fabrication by electrochemical and chemical dissolution. *Appl. Phys. Lett.*, **57** (10), 1046–1048.
8 Lehmann, V., and Gosele, U. (1991) Porous silicon formation: a quantum wire effect. *Appl. Phys. Lett.*, **58** (8), 856–858.
9 Brus, L. (1987) Size dependent development of band structure in semiconductor crystallites. *Nouv. J. Chim.*, **11** (2), 123.
10 Sailor, M.J. (1997) Sensor applications of porous silicon, in *Properties of Porous Silicon*, vol. 18, (ed. L. Canham), Institution of Engineering and Technology, London, pp. 364–370.
11 Greenwood, N.N., and Earnshaw, A. (1984) *Chemistry of the Elements*, Pergamon Press, Oxford.

12 Brinker, C.J., and Scherer, G.W. (1990) *Sol-Gel Science: The Physics and Chemistry of Sol-Gel Processing*, Academic Press, San Diego.

13 Canaria, C.A., Huang, M., Cho, Y., Heinrich, J.L., Lee, L.I., Shane, M.J., Smith, R.C., Sailor, M.J., and Miskelly, G.M. (2002) The effect of surfactants on the reactivity and photophysics of luminescent nanocrystalline porous silicon. *Adv. Funct. Mater.*, **12** (8), 495–500.

14 Bard, A.J., and Faulkner, L.R. (1980) *Electrochemical Methods*, John Wiley & Sons, New York, pp. 23–25.

15 Rouquerol, J., Avnir, D., Fairbridge, C.W., Everett, D.H., Haynes, J.H., Pernicone, N., Ramsay, J.D.F., Sing, K.S.W., and Unger, K.K. (1994) Recommendations for the characterization of porous solids. *Pure Appl. Chem.*, **66** (8), 1739–1758.

16 Gregg, S.J., and Sing, K.S.W. (1982) *Adsorption, Surface Area and Porosity*, 2nd edn, Academic Press Inc, London, p. 112.

17 Hérino, R., Bomchil, G., Barla, K., Bertrand, C., and Ginoux, J.L. (1987) Porosity and size distributions of porous silicon layers. *J. Electrochem. Soc.*, **134**, 1994–2000.

18 Collins, B.E., Dancil, K.-P., Abbi, G., and Sailor, M.J. (2002) Determining protein size using an electrochemically machined pore gradient in silicon. *Adv. Funct. Mater.*, **12** (3), 187–191.

19 Orosco, M.M., Pacholski, C., and Sailor, M.J. (2009) Real-time monitoring of enzyme activity in a mesoporous silicon double layer. *Nature Nanotech.*, **4**, 255–258.

20 Karlsson, L.M., Schubert, M., Ashkenov, N., and Arwin, H. (2004) Protein adsorption in porous silicon gradients monitored by spatially resolved spectroscopic ellipsometry. *Thin Solid Films*, **455–456**, 726–730.

21 Karlsson, L.M., Tengvall, P., Lundström, I., and Arwin, H. (2003) Penetration and loading of human serum albumin in porous silicon layers with different pore sizes and thicknesses. *J. Colloid Interface Sci.*, **266**, 40–47.

22 Bomchil, G., Halimaoui, A., and Hérino, R. (1989) Porous Si: the material and its applications to SOI technologies. *Appl. Surf. Sci.*, **41/42**, 604–613.

23 Zhang, X.G. (2004) Morphology and formation mechanisms of porous silicon. *J. Electrochem. Soc.*, **151** (1), C69–C80.

24 Weinberger, B.R., Peterson, G.G., Eschrich, T.C., and Krasinski, H.A. (1986) Surface-chemistry of Hf passivated silicon – X-ray photoelectron and ion-scattering spectroscopy results. *J. Appl. Phys.*, **60** (9), 3232–3234.

25 Chabal, Y.J., Chaban, E.E., and Christman, S.B. (1983) High-resolution infrared study of hydrogen chemisorbed on Si(100). *J. Electron. Spectrosc. Relat. Phenom.*, **29**, 35–40.

26 Yablonovitch, E., Allara, D.L., Chang, C.C., Gmitter, T., and Bright, T.B. (1986) Unusually low surface-recombination velocity on silicon and germanium surfaces. *Phys. Rev. Lett.*, **57** (2), 249–252.

27 Burrows, V.A., Chabal, Y.J., Higashi, G.S., Raghavachari, K., and Christman, S.B. (1988) Infrared-spectroscopy of Si(111) surfaces after HF treatment – hydrogen termination and surface-morphology. *Appl. Phys. Lett.*, **53** (11), 998–1000.

28 Chabal, Y.J. (1993) Infrared spectroscopy of semiconductor surfaces: H-terminated silicon surfaces. *J. Mol. Struct.*, **292**, 65–80.

29 Chabal, Y.J., Higashi, G.S., Raghavachari, K., and Burrows, V.A. (1989) Infrared-spectroscopy of Si(111) and Si(100) surfaces after HF treatment – hydrogen termination and surface-morphology. *J. Vac. Sci. Technol. A-Vac. Surf. Films*, **7** (3), 2104–2109.

30 Foll, H., Christopherson, M., Carstensen, J., and Haase, G. (2002) Formation and application of porous silicon. *Mater. Sci. Eng. R*, **39**, 93–141.

31 Christophersen, M., Langa, S., Carstensen, J., Tiginyanu, I.M., and Foll, H. (2003) A comparison of pores in silicon and pores in III-V compound materials. *Phys. Status Solidi A*, **197** (1), 197–203.

32 Hoffmann, R. (1987) How chemistry and physics meet in the solid-state. *Angew. Chem. Int. Ed. Engl.*, **26** (9), 846–878.

33 Ellis, A.B., Geselbracht, M.J., Johnson, B.J., Lisensky, G., and Robinson, W.R. (1993) *Teaching General Chemistry: A Materials Science Companion*, American Chemical Society, Washington, DC.

34 Kittel, C. (1986) *Introduction to Solid State Physics*, 6th edn, John Wiley & Sons, New York, pp. 291–299. Sze, S.M. (1981) *Physics of Semiconductor Devices*, John Wiley & Sons, New York.

35 Lehmann, V. (2002) *Electrochemistry of Silicon*, Wiley-VCH Verlag GmbH, Weinheim, pp. 51–75.

2
Preparation of Micro-, Meso-, and Macro-Porous Silicon Layers

This chapter sets up the basics for etching a single layer of porous silicon on either n- or p-type silicon using electrochemical or chemical etching methods. We describe the construction of the necessary apparatus, and provide basic procedures to prepare microporous (from p-type wafers), mesoporous (p^{++}-type wafers) and macroporous (n-type wafers) silicon. The important parameters to control in the electrochemical etching process are current density, electrolyte composition and homogeneity, light intensity (for n-type silicon), and temperature. We also cover two chemical methods to prepare porous silicon, where the oxidizing equivalents to drive the reaction are delivered by a chemical oxidant rather than a power supply. The first of these methods is stain etching, which is useful to make porous silicon from silicon powders or other forms of silicon that cannot conveniently be connected to an electrode. The second non-electrochemical method is metal-assisted etching, in which the pores are nucleated and propagate by means of metal nanoparticles previously deposited on the wafer surface. The oxidant in this case is hydrogen peroxide. Metal-assisted etching provides a convenient route to silicon nanowire arrays. More complicated porous structures (gradients, double layers, photonic crystals, and patterns spread over the surface) are described in Chapter 3.

2.1
Etch Cell: Materials and Construction

The etch cell is used to contain the electrochemical reaction. A general schematic diagram of a two-electrode cell was shown in Figure 1.2. There are many successful etch cell designs in the literature, and the user will have to decide which best suits his or her purposes. Appendix 1 provides the engineering diagrams for a few designs that will be used in the exercises in this book. The main design we will use will be referred to as the Standard etch cell. The nominal area of exposed silicon with this design is $1.2\,cm^2$, large enough for most research-scale experiments.

Porous Silicon in Practice: Preparation, Characterization and Applications, First Edition.
Michael J. Sailor.
© 2012 Wiley-VCH Verlag GmbH & Co. KGaA. Published 2012 by Wiley-VCH Verlag GmbH & Co. KGaA.

The material used to construct the etch cell should be composed of Teflon® or a similar material that is resistant to HF and organic solvents (ethanol primarily). O-rings are used to seal the cell, and these should be composed of either Kalrez® or Viton®. The Kalrez perfluoroelastomer is the most versatile of all the polymers used in O-rings. It is chemically resistant to polar solvents, organic solvents, inorganic and organic acids and bases, fuels, oils, lubricants, inorganic salts, aldehydes, metal halogen compounds, chlorine, caustic soda, aromatics, alcohols, and strong oxidizing agents. At elevated temperatures, the material can be attacked by diamines, nitric acid and basic phenols. Since some stain-etching procedures use nitric acid, the O-ring should be inspected and replaced if degradation is observed after a stain-etching procedure. The normal temperature service range for this O-ring is –40 to 315 °C.

An alternative fluoroelastomeric O-ring material that stands up well to aqueous HF/ethanol electrolytes is a copolymer of hexafluoropropylene and 1,1-difluoroethylene, known as Viton. It has excellent resistance to lubricants, mineral acids, non-polar compounds, oxidizing agents, halogenated aliphatic hydrocarbons such as carbon tetrachloride, and aromatic hydrocarbons such as toluene, benzene, or xylene. Viton is not recommended for basic and oxygenated solvents such as ammonia, ketones, ethers, esters, and hot anhydrous hydrofluoric, acetic, and chlorosulfuric acids. The normal temperature service range is from –23 to 204 °C. Kalrez and Viton O-rings can be obtained from many sources; Kimble-Kontes (www.kimble-kontes.com) is one of the more common.

The counter-electrode should be platinum, although stainless steel can be used if you can tolerate iron impurities in your samples and replacing the electrode occasionally. Platinum mesh is preferred (a nice mesh can be special ordered from Alfa Aesar, see Table 2.1), but a simple wire loop works too. If you want a uniform layer, it is important for the counter-electrode to be lying in a plane parallel to the silicon wafer, and fully immersed in the HF electrolyte.

2.2
Power Supply

Electrochemically etched porous silicon samples are usually prepared under controlled current conditions, meaning that the feedback loop in the power supply is set up to hold the current rather than the voltage constant. There are many inexpensive power supplies that can be used; a few are listed in Table 2.2. If you plan to etch photonic crystals or other multilayered structures, either one of the first three units in the list are acceptable. In general, the power supply should be able to deliver current in the range of a few mA up to at least 1 A. The higher end of this current range is used

Table 2.1 Electrodes and O-rings for etch cells.

Item	Vendor/cat #	Comments
Pt mesh counter-electrode	Special order from Alfa Aesar, (www.aesar.com). For Standard Cell, a custom platinum electrode with 15 mm mesh diameter, 100 mm length, 1 mm wire diameter, 1.5 mm mesh openings.	Mesh openings can sometimes trap H_2 bubbles during etch, leading to non-uniformity. If this becomes a problem, larger mesh openings, or an open loop or spiral of Pt wire (see next item) can be used.
Pt loop, Pt spiral counter-electrodes	Alfa Aesar (www.aesar.com): 10 cm length, 1 mm diameter Pt wire	You can fashion the piece of wire into a loop or a spiral. If you do not mind paying extra, you can have the vendor bend the spiral for you, if you specify the outer diameter (15 mm works well for the Standard etch cell).
O-rings for Standard etch cell (exposes 1.2 cm^2)	Kimble-Kontes (www.kimble-kontes.com): either Kalrez #112 O-ring: cat #758240-0112 or Viton #112 O-ring: cat #758252-0112	Viton is much cheaper (30×) than Kalrez and sufficient for most etching applications
	Top of the cell (seals the cell for anaerobic experiments) uses a Viton #110 O-ring: cat # 758252-0110	
O-rings for small etch cell (exposes 0.21 cm^2)	Kimble-Kontes (www.kimble-kontes.com): Viton #107 O-ring: cat # 758252-0107	
O-rings for large etch cell (exposes 8.6 cm^2)	Kimble-Kontes (www.kimble-kontes.com): TFE/Propylene #27 O-ring: cat # 758260-0027	

to produce high porosity, large pore diameters, or to electropolish or lift-off a pre-formed film. It is important to keep in mind that it is the current density, not the current, that determines parameters such as porosity, pore size, and whether or not the sample displays photoluminescence. The Standard etch cell used in this book (Appendix 1) exposes nominally 1.2 cm^2 of silicon to the etching solution, so the current density needed (current divided by area) can be delivered fairly easily by most systems. However, if

Table 2.2 Examples of power supplies that can be used to etch porous Si.

Model	Vendor	Comments
ATE 25-2DM	Kepco (www.kepco.com)	Great etching tool based on an op-amp controlled current source. Low noise, linear response, can deliver up to 2 A. The digital control by GPIB interface (via National Instruments Labview program) is too slow to handle high data rates needed to prepare rugate or other photonic crystal structures as of this writing (a driver to handle buffered i/o had not been implemented). However, the well-behaved analog input (specify the analog input option when ordering) is fast enough (<1 ms) to follow the output from a computer-controlled DAC card, providing great versatility for many user-defined experiments. A bipolar version is also available as the BOP-50-4D (allows you to swing positive and negative of zero, not necessary for etching).
6612C	Agilent (www.agilent.com)	40 Watt System Power Supply. Can program simple waveforms from front panel, or more complicated ones by computer (GPIB) interface. The interface handles data rates up to 250 points s^{-1}.
2601 Sourcemeter	Keithley (www.keithley.com)	Can program simple waveforms from front panel, or more complicated ones by computer interface. As of this writing the interface handles data rates up to 500 points s^{-1}, adequate to prepare rugate or other photonic crystal structures – the version we tested produces short current spikes when changing scales or settings so it should be used with the automatic current range adjustment disabled.

PAR 263A/94	Princeton applied research (www.princetonappliedresearch.com)	General purpose electrochemical potentiostat/galvanostat. This system has an external analog input that allows the user to supply a voltage signal from a computer-controlled DAC card, useful for etching photonic crystals such as rugates, Bragg stacks, etc. The /94 designation is for a current amplifier, to allow the unit to run currents up to 2 A.
P6000 programmable power supply	Protek (www.protektest.com)	Inexpensive power supply that can operate in constant current (up to 3 A) or constant voltage (up to 30 V) mode. It can be controlled with an external computer, but cannot output a rapidly changing arbitrary current–time waveform needed for more elaborate photonic crystal structures. Good for simple, single layer etches.
PARSTAT® 2273 potentiostat/ galvanostat	Princeton applied research (www.princetonappliedresearch.com)	General purpose electrochemical potentiostat/galvanostat. The "Cadillac" of electrochemical instruments. It is designed for a wide range of electrochemical experiments. It cannot output the type of user-defined current–time waveforms needed to etch photonic crystals. Like any galvanostat, it can output a constant current useful for etching single-layer structures.

it is desired to etch an entire 8 in wafer, the power supply will have to deliver 16 A in order to etch at a modest current density of 50 mA cm^{-2}! These higher currents necessitate extra precautions such as thicker gauge wires and active cooling of the apparatus, not to mention the handling of fairly large volumes of HF solution. If only single porous layers are to be prepared (such as Fabry–Perot films or luminescent material), a constant current source is sufficient. The samples that require more complicated waveforms, such as photonic crystals, need a computer interfaced to the power supply (see Chapter 3).

2.3
Other Supplies

A list of tools and supplies useful to have on hand is given in Table 2.3. Some of these are a matter of personal preference.

2.4
Safety Precautions and Handling of Waste

Hydrofluoric acid is toxic and can cause severe chemical burns that are slow to heal. In the case of exposure, the affected area should be thoroughly rinsed with water, and treated with HF antidote gel; medical care should then be sought immediately (see Appendix 2 for more details).

Personal protection. Etching and all other handling of HF should be performed in a fume hood, the hood windows should be used as additional splash protection between the work and your face and neck. Lab coats or a long apron and closed shoes should be worn, along with chemical splash goggles for eye protection. Gloves (nitrile inner gloves and neoprene outer gloves) should be used, and they should be removed as soon as the HF operation is finished. The inner gloves should be disposed of in an appropriate waste container.

Handling of waste. Used HF solutions should be treated as hazardous waste. Many labs prefer to neutralize their HF solutions prior to delivery to the waste disposal personnel. This also reduces the amount of HF gas that escapes into the environment. Neutralization can be accomplished with $CaCO_{3(s)}$. (Item #36337, Alfa Aesar, www.alfa.com) Fill a 2 l Nalgene™ beaker or similar container one quarter full of $CaCO_3$ powder and keep it in the back of the fume hood in which you are handling the HF. The used etching solutions can be transferred directly from the etch cell into this beaker. Dispose of it when the beaker is half full.

Table 2.3 Supplies for etching and handling porous silicon.

Item	Vendor/cat #	Comments
Teflon tweezers		Unless you want to etch a replica of your fingerprint into a piece of porous Si, you should always handle your samples with Teflon or other plastic tweezers. Metal tweezers can scratch the surface and they can also be corroded by traces of residual HF.
HF antidote gel	Pharmascience Inc. 8400 Darnley Rd. Montreal, Canada. H4T 1M4 Phone: (514) 340-1114 Fax: (514) 342-7764 U.S. (Buffalo, NY) distributor: 1-800-207-4477	This is discussed in more detail in Appendix 2.
$CaCO_{3(s)}$	Alfa Aesar, www.alfa.com Item #36337	Keep a bucket of this on hand to dispose of old HF solutions and to neutralize HF spills.
Ultrasonic bath	VWR® (www.vwr.com) Model 50T ultrasonic cleaner, part # 21811-820	For cleaning silicon wafers prior to etch; for preparing micro- or nanoparticles from porous silicon films.
Nalgene containers: 2 l beaker, 100 ml graduated cylinder, 250 ml storage bottles		You cannot store HF solutions in glass containers.
Neoprene gloves		To wear when handling HF solutions. You can also wear nitrile or latex gloves under these.
Safety glasses/ goggles		You should always wear eye protection when handling HF solutions.
Ethanol wash bottle		For rinsing samples
Nitrogen spray gun	Innotech Products (www.innotechprod.com) TA-N2-1000	To blow samples dry after rinsing

(Continued)

Table 2.3 (Continued)

Item	Vendor/cat #	Comments
Analytical Balance	Sartorius CP225D www.sartorius-mechatronics.com	For gravimetric determination of porosity. For the 1.2 cm^2 samples prepared using the Standard etch cell (Appendix 1), you should use a balance with 0.01 mg precision.
Plastic secondary containment tray		It is a good idea to place the etching cell in a plastic tray in case of accidental spills or a leaky cell
Plastic pipette droppers		For transferring HF electrolyte into the etching cell
Anti-static gun	Sigma Aldrich chemicals (www.sigmaaldrich.com) Zerostat anti-static instrument, part # Z108812-1EA	For discharging samples before weighing when performing gravimetric porosity measurement
Light power meter:	Coherent, inc. (www.coherentinc.com) Fieldmate power meter, part # 1098297 with PS10Q sensor, part # 0012-4600	Useful for measuring light intensity used in photoelectrochemical etches

2.5
Preparing HF Electrolyte Solutions

Hydrofluoric acid dissolves glass, and so all solutions need to be prepared and stored in plastic bottles. In the experiments described in this book we use a mixture of aqueous HF with ethanol unless otherwise stated. Hydrogen-terminated porous silicon is fairly hydrophobic, and water does not wet the pores very well. The ethanol is there primarily to reduce the surface tension of the electrolyte, allowing it to enter the very small pores as they are formed. Other alcohols or water-miscible organic solvents can be used, though they can exert a profound influence on the pore morphology.

The basic procedure to mix a solution involves volume dilution by pouring the aqueous HF solution into a graduated cylinder and then diluting it with absolute (200-proof) ethanol. The resulting solution is then stored in a plastic bottle with a cap. Semiconductor-grade aqueous HF can be purchased from various sources (Fisher Scientific Sigma-Aldrich), and it comes as a 48% or 49% (by mass) aqueous solution. For the 3:1 HF:ethanol solution used in Experiment 2.1, fill the graduated cylinder with 30 ml of aqueous 49% HF and dilute it to the 40 ml mark (i.e., add ~10 ml) with absolute ethanol. A typical preparation uses about 3 ml of this solution, and the electrolyte should be disposed of after each etch.

2.6
Cleaning Wafers Prior to Etching

Although they are polished to a mirror finish and look very clean when you pull them from the manufacturer's container, silicon wafers contain impurities and oxide on their surface that must be removed prior to etching in order to obtain consistent results. In particular, optical devices (e.g., 1-d photonic crystals) require a robust cleaning procedure to provide the same optical spectrum from sample to sample. The following procedures are useful – the specific method will depend on the degree of cleanliness your application requires. The method preferred by the author's lab is that described in Section 2.6.4.

2.6.1
No Precleaning

The etching solution and the electrochemical etching process remove oxides and many surface impurities. For simple demonstration samples, no prior cleaning is necessary and the silicon wafer can be used "as is".

2.6.2
Ultrasonic Cleaning

Ultrasonic cleaning removes oil and other organics from the silicon wafer, but not the surface oxide. A quick 15 s dip in ethanolic HF (the etching solution described above) will remove the native surface oxide and some organic residues:

1) Holding the sample with Teflon® tweezers, rinse the sample in ethanolic HF for 15 s.
2) Rinse sample in a stream of pure ethanol, blow dry in a stream of pure nitrogen.
3) Sonicate successively in chloroform ($CHCl_3$), acetone and ethanol for 10 min in each solvent.

Be sure to handle the sample with Teflon tweezers after cleaning.

2.6.3
RCA Cleaning

Very thorough cleaning can be accomplished using the reactive chemical etches developed by the semiconductor industry to prepare wafers for microelectronics fabrication. The "RCA" etch procedure given below is the standard. This is a good procedure for preparing samples that need to be atomically smooth (for AFM analysis of surface topology, deposition of lipid monolayers, etc.):

1) Boil the sample in ethanol for 30 min.
2) Sonicate successively in chloroform ($CHCl_3$), acetone and ethanol for 10 min in each solvent.
3) Soak for 15 min in RCA1 ($H_2O:NH_4OH:H_2O_2$ 5:1:1) at 75 °C
4) Soak for 15 min in RCA2 ($H_2O:HCl:H_2O_2$ 5:1:1) at 75 °C
5) Rinse with deionized water and then ethanol

Note the H_2O_2 used in the above procedure must be at a concentration of 30% in water. As a point of reference, the concentration of H_2O_2 in the hydrogen peroxide solutions sold in pharmacies as a disinfectant is 3%. While not quite rocket fuel, the 30% H_2O_2 used in the above preparation is pretty hazardous. It will react very quickly if spilled on the skin, leaving white blisters. The HCl and NH_4OH used in the above preparation refer to the concentrated aqueous reagents, 37% and 29% by weight, respectively.

2.6.4
Removal of a Sacrificial Porous Layer with Strong Base

This method takes advantage of the fact that porous silicon dissolves in strongly basic solutions. It is a relatively quick and thorough cleaning procedure that provides good, reproducible surfaces. It is particularly useful for etching p^{++} samples for optical applications. The procedure is described in detail in Experiment 2.2:

1) Sonicate the sample in isopropyl alcohol for 10 min.
2) Etch the sample in 3:1 aqueous 48% HF:ethanol at 200 mA cm^{-2} for 30 s.
3) Soak the sample in 1 M aqueous NaOH (or KOH) solution containing 10% ethanol for 5 min.

After this procedure, you can rinse the cell out, add the HF electrolyte solution and then proceed with your porous silicon etch. For the belt-and-suspenders enthusiasts, the sample can be cleaned with an RCA process prior to this procedure.

2.7
Preparation of Microporous Silicon from a p-Type Wafer

Typical average pore sizes for porous silicon prepared from p-type substrates are <2 nm. The pores overlap to the extent that the silicon nanostructures remaining in the film take on the form of very small (~5 nm) pillars, or nanowires. (Figure 2.1). This type of structure has been used to make conductivity-based sensors, luminescent nanostructures, and readily dissolved sacrificial release layers. It is also the type of material that first displayed visible photoluminescence due to quantum confinement effects [1, 2].

Since this is the first etching experiment in this book, more detail will be given than in subsequent experiments. We will also demonstrate how to calculate the reaction stoichiometry of the etching process in this experiment.

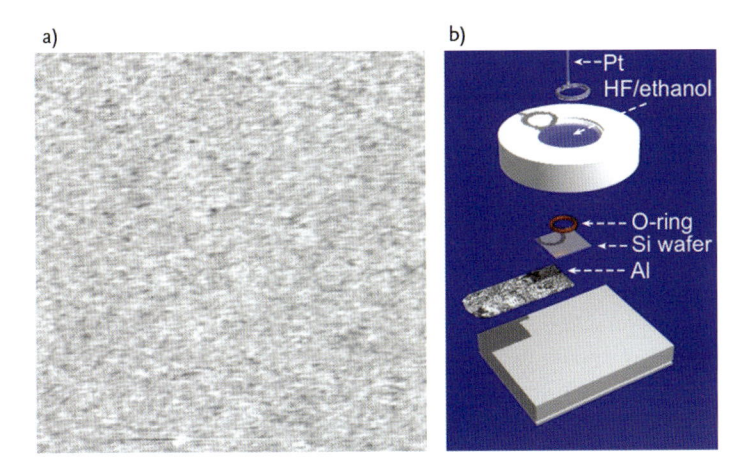

Figure 2.1
(a) A plan-view atomic force microscope image of the surface of a porous silicon sample prepared from p-type silicon. This sample was prepared from a 3 Ω cm resistivity, (100) polished wafer, etched in a 1:1 (by volume) solution of 49% aqueous HF:ethanol at a constant current density of 30 mA cm^{-2} for 5 min. The image frame is 200 × 200 nm^2; the pore diameters in this sample are of the order of just a few nm. (b) An exploded view of the standard electrochemical etch cell used in this experiment, showing the placement of the counter-electrode, the silicon sample, and the aluminum foil back-contact.

EXPERIMENT 2.1: Electrochemical etch of microporous silicon from p-type silicon

In this experiment you will prepare a porous silicon sample from p-type silicon. You will also calculate the current efficiency of the etch.

Equipment/supplies:

Constant current power supply	See Table 2.2 for examples. For this example, a Princeton Applied Research PAR 363 potentiostat/galvanostat, operating in galvanostat mode, was used.
Pt loop counter electrode	See Table 2.1. You can also use a Pt mesh for this experiment
Al foil back-contact	This should be fabricated from a heavier gauge aluminum than is typically sold as household aluminum foil – aluminum weighing boats from Alfa Aesar (www.alfa.com) can be cut to size
Etch cell	Standard etch cell, Appendix 1
p-type Si chip, (100), 0.5 Ω cm	Siltronix (www.siltronix.com). Measurement of resistivity is performed in Experiment 1.1
1:1 aqueous HF (48%): ethanol solution	See Section 2.5

Procedure:

Because of the low resistivity of the parent wafer, the resistance between it and our aluminum foil back side contact is significant. Therefore, an ohmic contact should be placed on the back side of the chip before etching to provide a more uniform and reproducible etch. An ohmic contact can be made by evaporating a thin layer of aluminum on the back side of the wafer, as described in Chapter 1 (Section 1.8.4.1).

1) Obtain a boron-doped, p-type, polished (100) silicon wafer, with a resistivity of approximately 0.5 Ω cm, cut to 2 × 2 cm^2. Soak or rinse the front side of the sample in the HF etchant solution for 15 s, rinse it with ethanol, dry it, and weigh it to the nearest 0.01 mg.

2) Mount the chip in the standard Teflon etch cell (Appendix 1), using a piece of aluminum foil as a back-contact and a Viton O-ring to seal the cell. Mount the aluminum on the Teflon base, and tape it in

2.7 Preparation of Microporous Silicon from a p-Type Wafer

place with a small piece of tape. Make sure the tape does not cover the area where your silicon wafer will make contact with the foil.

3) After gently tightening the screws by hand, tighten each screw one quarter turn at a time so as to apply even pressure. If the wafer cracks, you will need to replace it. Once the screws are tightened test the cell for leaking by placing ethanol in the cell and allowing it to sit for a few minutes.

4) Place your cell in the fume hood designated for handling HF and add the 1:1 (v/v) mixture of 48% aqueous HF and absolute ethanol. The electrolyte level should be a few mm below the top of the cell, the volume of approximately two droppers.

5) Immerse the loop of platinum wire (counter-electrode) in the electrolyte. Make sure the loop is fully submerged. Attach the counter-electrode to the negative (black) lead of the power supply. Attach the positive (red) lead to the aluminum foil back-contact.

6) Set up the power supply in constant current mode (galvanostat) and make sure that it is set to deliver a constant current density of $30\,\text{mA}\,\text{cm}^{-2}$, corresponding to a measured current of 36 mA if using the Standard etch cell (Appendix 1). If you are unfamiliar with your system, it is a good idea to set the etching cell up in series with an ammeter so that you can monitor the current passing through the cell in real time (Figure 2.2 shows a wiring diagram). Most power supplies have a meter or an auxiliary output that will allow you to do this, though it may not have the resolution (± 1 mA at most, ± 0.01 mA preferred) needed to monitor current precisely.

Note it is *current density* (current passed per unit area of silicon exposed to the solution) that is the important parameter here, not the absolute current. Different researchers use different cells, and you should always report etch current in terms of current density to ensure reproducibility from lab to lab. In the present case, the Standard etch cell (Appendix 1) exposes $1.2\,\text{cm}^2$ of silicon to the electrolyte solution, so the current value on the galvanostat should be set to $30\,\text{mA}\,\text{cm}^{-2} \times 1.2\,\text{cm}^2 = 36\,\text{mA}$.

7) Activate the power supply and apply the anodic current between the aluminum back-contact (positive lead) and the platinum counter-electrode (negative lead) for 10 min. Record the exact current and time (s) used.

8) After completion of the etch, turn off the power source and remove the HF solution from the cell using a plastic dropper. Rinse the cell three times with ethanol and discard the washings in a designated waste container. Carefully pick up the cell and rinse the face of the cell and the chip with ethanol into the waste container. Take the cell

apart and rinse the chip again with ethanol, then blow it dry under a stream of nitrogen. Thoroughly rinse the pieces of the cell and the O-ring with ethanol. This procedure should be performed in the same hood in which you perform the etch, and you should wear protective gloves throughout. Weigh the sample after it is dry.

Results:

To calculate the current efficiency of the etch, you must convert current to coulombs and mass to moles. Faraday's constant is the conversion factor relating coulombs to moles of electrons, and it has a value of 96 485 C mol^{-1}. Table 2.4 presents the results for a representative sample.

The reaction stoichiometry, η, can be represented as the number of moles of electrons passed in the circuit (n_e) per mole of silicon atoms removed (n_{Si}). For a current i amperes, passed for t seconds:

$$\eta_{curr} = \frac{it}{Fn_{Si}} \qquad (2.1)$$

Where F is Faraday's constant. In the present case this works out to (note that one mA is 1 mC s^{-1}):

$$\frac{36\,mC}{sec} \times \frac{1C}{1000\,mC} \times \frac{600\,sec}{96,485\,C} \times \frac{1\,mol\,e^-}{0.00297\,g\,Si} \times \frac{28.086\,g\,Si}{1\,mol\,Si}$$

$$=2.12$$

Thus two electrons pass through the circuit for every atom of silicon removed. This works out pretty close to the 2:1 electrons per silicon atom stoichiometry of Equation 1.9. If we assume that is the correct stoichiometry, 0.12 extra electrons passed through the circuit for each silicon atom removed. The extra electrons observed in the above experiment can probably be attributed to weighing errors, though parasitic reactions (e.g., water electrolysis) or the 4-electron silicon oxidation reaction (Equation 1.6) could also be responsible.

Note that the alligator clips in the etching apparatus will corrode with time, degrading the contacts and increasing the total resistance of the circuit. The constant current source will compensate by increasing the applied voltage in order to maintain the current set point. It is a good idea to monitor and record the voltage applied by the source during an etch. If the applied voltage is too large (typically >15 V), it is an indication of faulty connections – check or change the alligator clips or the aluminum back-contact. A sudden jump in voltage during an etch is often a sign of a leaky cell – the HF electrolyte has drained, breaking contact with the Pt counter-electrode or attacking the aluminum back contact. In this case be cautious in disassembling the cell because it will be contaminated with HF solution.

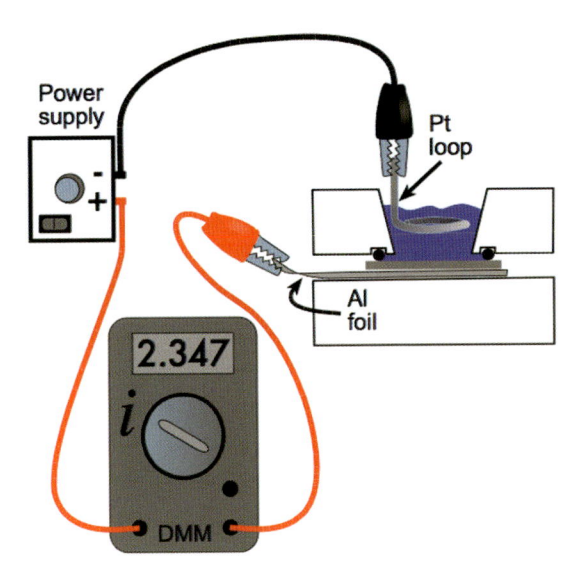

Figure 2.2
Wiring diagram used to monitor current during an electrochemical etch. The digital multimeter (DMM) is set to measure current (i), and the circuit is connected in series. Be sure you do not exceed the rated current setting of the DMM or you will blow its internal fuse.

Table 2.4 Data used to determine stoichiometry of the electrochemical etch.

Current, mA	36
Time, s	600
Mass Si chip before etch, g	0.27722
Mass Si chip after etch, g	0.27425
Mass Si removed in etch, g	0.00297

2.8
Preparation of Mesoporous Silicon from a p^{++}-Type Wafer

Mesoporous silicon is readily fabricated from highly doped, p^+- or p^{++}-type silicon. The "+" superscripts on p^+ and p^{++} indicate that these samples are highly doped with the p-type dopant (typically boron). The resistivity of p^{++} silicon is so low (~0.001 Ω cm) that it has almost metallic conductivity, and it is referred to as degenerately doped material. The lower resistivity of this semiconductor relative to p-type silicon causes a change in the etching mechanism. Whereas the morphology of p-type silicon is determined by quantum confinement and electric field-driven effects in the silicon nanowires, the etching mechanism for p^{++} silicon is dominated by carrier tunneling [3]. The result is fewer pores with larger diameters. The pore

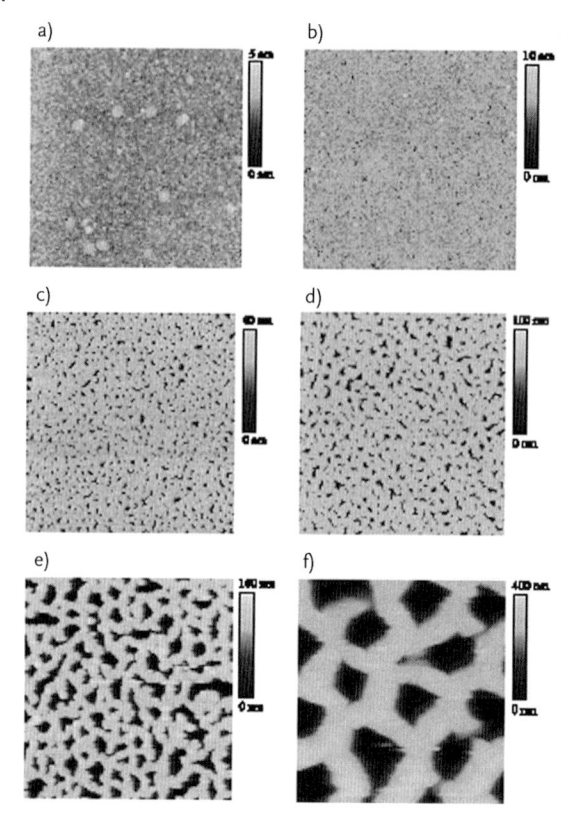

Figure 2.3
Control of pore size with current in highly doped p-type silicon. Plan-view atomic force microscope images of six mesoporous silicon samples, etched from six p⁺-type wafers having a resistivity of 1.0 mΩ cm. The electrolyte used was 3:1 (by volume) 49% aqueous HF:ethanol. The image size and current densities used for the samples shown: (a) $1.5 \times 1.5\,\mu m^2$, $150\,mA\,cm^{-2}$; (b) $5 \times 5\,\mu m^2$, $295\,mA\,cm^{-2}$; (c) $5 \times 5\,\mu m^2$, $370\,mA\,cm^{-2}$; (d) $5 \times 5\,\mu m^2$, $440\,mA\,cm^{-2}$; (e) $5 \times 5\,\mu m^2$, $515\,mA\,cm^{-2}$; (f) $5 \times 5\,\mu m^2$, $600\,mA\,cm^{-2}$. All samples were etched to a constant number of coulombs of $4.5\,C\,cm^{-2}$. Taken from [7].

diameters are determined by the value of the applied current, and they generally fall in the range 10–200 nm. Many proteins, enzymes, and antibodies have dimensions of the order of 10 nm, and mesoporous silicon has found use in many biotechnology-related applications including biosensors [4–6]. Examples of the range of pore sizes accessible just by changing the current density used in the etch are given in Figure 2.3.

The current density we use in this experiment is rather large, so we will use the platinum coil counter-electrode. A platinum mesh will also work (Table 2.2). You can use a platinum loop, but the lower surface area of platinum exposed to the solution will force the potentiostat to apply higher voltages to drive the reaction, which may affect the reproduciblity of the etch. It is good practice in electrochemistry to use a counter-electrode that has a surface area larger than the working electrode, so that the current is limited by the fundamental electrochemical process of the reaction taking place at the working electrode (i.e., the reaction you care about), rather than mass transfer effects at the counter-electrode.

The current density used in this experiment is close to the electropolishing limit, which is highly dependent on the sample resistivity. If your sample comes out of the etching bath displaying cracks and flakes, or if there is no porous silicon at all, you should reduce the current density by 5% and try again.

Etch of the highly doped wafers used in this experiment can generate a top "crust" of porous material with pores smaller than the rest of the film. This phenomenon appears to be related to segregation of dopants at the surface of the wafer. Therefore, a preliminary electropolishing step is performed in this experiment to remove the top few nm of silicon from the wafer surface prior to etching the porous layer. Examples of surface pore morphologies resulting from pre-cleaning by electropolishing are given in Figure 2.4. The method is referred to earlier in Section 2.6.4

2.9
Preparation of Macroporous, Luminescent Porous Silicon from an n-Type Wafer (Frontside Illumination)

Luminescent porous silicon can be prepared from either n- or p-type substrates. In the experiment in this section we will use an n-type wafer. In order to etch n-type silicon, the sample must be illuminated to supply the current needed. This is because the diode-like behavior of the silicon electrode results in a reverse bias situation relative to the corrosion current, as described in Chapter 1 (Section 1.7.10). Samples can be illuminated either from the front or from the back, and the next two experiments provide examples of each. There are distinct differences in morphology that result from the two experimental configurations.

The samples resulting from Experiments 2.3 and 2.4 will generally display a fairly intense orange photoluminescence right out of the etch bath. The light emission comes from recombination of quantum confined electron–hole pairs in nanocrystalline silicon domains, formed during the pore etching process. When the pores are numerous enough to overlap significantly, very small features become isolated from the bulk silicon substrate. These isolated structures form an ensemble of interconnected nanometer-sized silicon crystallites, whose dimensions are small enough to exhibit quantum confinement effects [1, 8]. Photoexcitation with blue or UV light generates confined electron–hole pairs in the nanocrystallites of

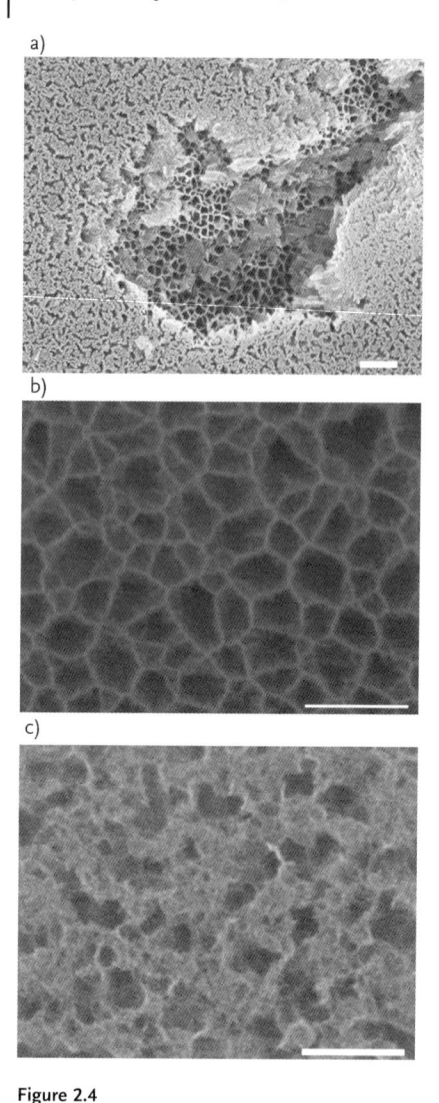

Figure 2.4
Plan-view images of porous silicon surfaces, showing the effect of surface cleaning on pore morphology. (a) A "crust" layer of micro/mesoporous silicon is often generated when p^{++} type silicon is etched. This image shows a sample with the crust partially removed, revealing the mesoporous layer beneath. This crust is not observed if the wafer is cleaned appropriately prior to the etch. (b) Sample pre-cleaned using the method described in Section 2.6.4 displays well-defined mesopores without a surface crust layer. (c) Electropolishing the wafer prior to etching may not be as effective at removing the crust. Both samples from (b) and (c) were etched from the same silicon wafer, using the same electrolyte composition (3:1 v:v aqueous 48% HF:ethanol), the same current density (540 mA cm^{-2}), and the same etch time (35 s). The samples only differed by the pre-etch cleaning procedure used. Scale bar in all images is 200 nm. Images courtesy Frederique Cunin and Michelle Chen.

EXPERIMENT 2.2: Electrochemical etch of mesoporous silicon from p^{++}-type silicon

In this experiment you will prepare a porous silicon sample from p^{++}-type silicon. We will also preclean the wafer with a sacrificial porous layer to improve reproducibility. This experiment is also used to demonstrate gravimetric measurements for the determination of thickness and porosity in Experiment 5.1.

Equipment/supplies:

Constant current power supply	See Table 2.2 for examples. For this example, a Princeton Applied Research PAR 363 potentiostat/galvanostat, operating in galvanostatic mode, was used.
Pt spiral counter-electrode	See Table 2.1. You can also use a Pt mesh or loop for this experiment
Al foil back-contact	Aluminum weighing boats from Alfa Aesar (www.alfa.com) can be cut to size
Etch cell	Standard etch cell, Appendix 1
p^{++}-type (100) Si chip, resistivity 1 mΩ cm	Siltronix (www.siltronix.com). Measurement of resistivity is performed in Experiment 1.1
1 M NaOH in 9 water:ethanol	This solution is used to remove the sacrificial layer of porous silicon
3:1 aqueous HF (48%): ethanol solution	See Section 2.5

Procedure:

1) Obtain a boron-doped, p^{++}-type, polished (100) silicon wafer, with a resistivity of approximately 1 mΩ cm, cut to 2 × 2 cm^2.

2) Clean the sample in an ultrasonic bath containing isopropyl alcohol for 15 min.

3) Mount the chip in the standard Teflon etch cell (Appendix 1), using a piece of aluminum foil as a back-contact to the silicon.

4) Place your cell in the fume hood designed to handle HF and add the 3:1 (v/v) mixture of 48% aqueous HF and absolute ethanol.

5) Immerse the platinum wire counter-electrode in the electrolyte. Attach the counter-electrode to the negative (black) lead of the power supply. Attach the positive (red) lead to the aluminum foil back-contact.

6) Set the power supply to constant current mode (galvanostat) to deliver a current density of 200 mA cm^{-2} (corresponding to a measured current of 240 mA if the Standard etch cell is used).

7) Activate the power supply and apply anodic current between the aluminum back-contact (positive lead) and the platinum counter electrode (negative lead) for 30 s. This step creates several nm of a sacrificial porous layer that you will remove in the next step.

8) Remove the HF electrolyte and fill the cell with the NaOH solution. You should notice immediate and vigorous bubbling as the porous layer dissolves and releases H_2. The bubbling should cease within a few minutes.

9) After 5 minutes, remove the NaOH solution and rinse the cell thoroughly with ethanol.

10) If you plan to perform the destructive gravimetric measurement of porosity (Experiment 5.1), remove the sample from the cell, rinse it with ethanol, dry it, weigh it to the nearest 0.01 mg, and remount the sample in the etch cell. Otherwise, continue with the next step:

11) Add electrolyte with 3:1 (v/v) mixture of 48% aqueous HF and absolute ethanol and immerse the counter-electrode.

12) Set the power supply (in galvanostatic mode) to deliver a constant current density of 420 mA cm^{-2} (corresponding to a current of 504 mA if the Standard etch cell is used).

13) Activate the power supply and apply the anodic current for 60 s. Record the exact current and time (in s) used.

14) After completion of the etch, turn off the power source and remove the HF solution from the cell. Rinse the cell with ethanol, remove the chip and rinse it again with ethanol, then blow it dry under a stream of nitrogen. Weigh the sample after it is dry.

Results:

This sample will be ~12 μm thick, with pores of the order of 50 nm and a porosity of ~75%. These samples will display optical interference effects, and faint concentric lines (Newton's rings) can be observed on the surface of the sample under the right lighting. Experiment 5.2 describes measurement of the optical spectrum of the film.

porous silicon [9]. The energy gap of the nanocrystallites is larger than the bulk silicon bandgap (1.1 eV, corresponding to the near-infrared region of the spectrum) by the confinement energy and a visible red, orange, or green photoluminescence (PL) is observed upon recombination. The preparation described below generates a material with a bright orange photoluminescence that is easily observed upon excitation with a small UV LED. Note that, although the material appears by SEM to be macroporous, there are many microporous regions that yield the fine silicon nanostructures needed for quantum confinement.

A major requirement when photoetching n-type silicon is to maintain sufficient light intensity to support the desired etching current. As discussed in Chapter 1, the hole current needed to corrode the silicon wafer is blocked by the rectifying property of the n-silicon/electrolyte interface. This was not a problem when etching the p-type wafer in Experiment 2.1, because the p-silicon/electrolyte junction is in forward bias during etch and the corrosion current increases more or less exponentially with increasing applied voltage. Current blocking was also not an issue with the highly doped p^{++} sample etched in Experiment 2.2, because that type of silicon is so highly doped it is essentially metallic. However, the n-silicon/electrolyte interface is in reverse bias, and the hole current needed to corrode the wafer must be generated by photoexcitation.

2.9.1
Power Supply Limitations

It is important to pay attention to the absolute voltage being applied to the cell when performing a constant current etch. The feedback loop imposed by the constant current source follows the current–voltage relationship of Ohm's law:

$$V = iR \tag{2.2}$$

In this case, V is the voltage applied across the cell, i is the current setpoint determined by the user, and R is the total series resistance of the circuit, including the electrochemical cell, the electrode contacts, and the leads. The feedback loop increases the output voltage until the current sensed in the etching circuit matches the setpoint current. All power supplies have an upper limit to the voltage they can apply, known as the compliance voltage. If the resistance of the cell is too large, the device will reach its compliance voltage limit and the current flowing through the cell will be less than the programmed value.

As an example, let us assume we want to etch a sample (either n- or p-type) at a constant current density of 75 mA cm^{-2} using the Standard etch cell (Appendix 1) and a Princeton Applied Research PAR 363 galvanostat. The PAR 363 has a compliance voltage of 12.5 V. The Standard etch cell exposes

1.2 cm² of silicon, so the PAR 363 must source $75 \times 1.2 = 90$ mA. If the resistance of the cell is 200 Ω, the voltage required to maintain a current of 90 mA is $iR = 200 \times 0.09) = 18$ V. This exceeds the compliance voltage of the instrument, so the power supply will go into an overload condition. The PAR 363, and many other commercial power supplies, incorporate an "amplifier overload" or "control amplifier overload" indicator on their front panel to signal when the feedback loop is demanding more voltage than can be supplied. You must be aware of this potential situation, because the actual current in the circuit may be significantly lower than the value you set. In the example above, the sample would be etching at only $12.5/200 = 0.0625$ A (62.5 mA), corresponding to a current density of 52.1 mA cm^{-2}. This is a third lower than the desired value of 75 mA cm^{-2}. Remember that pore morphology and porosity are highly dependent on current density.

A current amplifier overload condition is most often encountered when etching n-type silicon, because the material requires a lot of light to support even moderate currents. In addition, irreproducible results are often encountered if the light intensity is barely sufficient, requiring the power supply to apply voltages of 6 V or larger. In this situation the voltage supplied by the source can fluctuate wildly, in particular when bubbles formed during the etch create shadows on the wafer surface (if using front side illumination). The best way to avoid this problem is to keep the light intensity high, and to minimize the contact resistance everywhere else in the circuit: make sure the alligator clips and aluminum foil back-contact are clean and uncorroded. A brief HF rinse to remove the native oxide from the back side of the silicon wafer before mounting it in the cell lowers the contact resistance between the aluminum foil and the silicon wafer. If the Al/Si contact resistance is still too high, you will need to make an ohmic contact to the back side of the silicon wafer (see Section 1.8.4).

You should also monitor the voltage being supplied by the power source throughout the etch. Most constant current sources have a display on the front panel or an output jack that allows the user to directly monitor the applied voltage. If that is not available, Figure 2.5 gives a schematic showing how to monitor the applied voltage drop across the etch cell with any power supply. For a typical etch using the Standard etch cell (Appendix 1), the applied voltage should be in the range 1–5 V. If the voltage across the cell is too high it is an indication that the illumination lamp is not bright enough or the aluminum foil contact is too resistive.

2.10

Preparation of Macroporous, Luminescent Porous Silicon from an n-Type Wafer (Back Side Illumination)

Back side illumination of n-type silicon provides a means to make uniform-diameter macropores, and avoids shadowing effects of the counter-electrode

2.10 Preparation of Macroporous, Luminescent Porous Silicon

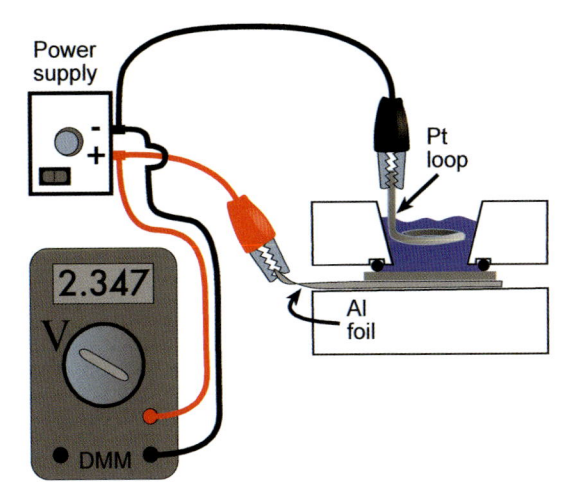

Figure 2.5
Schematic depicting the connections used to monitor the voltage drop across an etch cell during an experiment. The voltage should be fairly constant and low relative to the compliance voltage of the power supply throughout the process when photoetching an n-type silicon wafer.

EXPERIMENT 2.3: Photoelectrochemical etch of an n-type silicon wafer by front side illumination

In this experiment you will prepare a luminescent porous silicon sample from an n-type wafer. This procedure produces a macroporous sample that is ~20 μm thick, with a porosity of ~70%

Equipment/supplies:

Constant current power supply	See Table 2.2 for examples. For this example, a Princeton Applied Research PAR 363 potentiostat/galvanostat was used, operating in galvanostatic mode
Pt loop counter-electrode	See Table 2.1. The loop is needed because you must illuminate the sample from the front side without shadows
Al foil back-contact	Aluminum weighing boats from Alfa Aesar (www.alfa.com) can be cut to size
Etch cell	Standard etch cell, Appendix 1
n-type (100) Si chip, resistivity 0.8 Ω cm	Siltronix (www.siltronix.com). Measurement of resistivity is performed in Experiment 1.1

1:1 aqueous HF (48%): ethanol solution	See Section 2.5
Light source	Dolan-Jenner MI-150 high intensity fiber illuminator (www.dolan-jenner.com). A simple high intensity tungsten halogen desk lamp can also be used for front side illumination.
UV LED illuminator	In order to view the photoluminescence. Purple Freedom Micro (blacklight), or UV Freedom Micro ("pure" UV), both from Photon Light (www.photonlight.com).

Procedure:

1) Rinse the n-type silicon chip with HF/ethanol and dry it.

2) Mount the chip in the standard Teflon etch cell (Appendix 1), using a piece of aluminum foil as a back-contact to the silicon.

3) Place the etch cell in the fume hood and add the 1:1 (v/v) mixture of 48% aqueous HF and absolute ethanol.

4) Immerse the platinum wire loop counter-electrode in the electrolyte. Attach the counter electrode to the negative (black) lead of the power supply. Attach the positive (red) lead to the aluminum foil back-contact. If you are using an electrochemical power supply such at the Princeton Applied Research PAR 363, the reference and counter electrode leads must be shorted together and connected to the platinum loop; the working electrode lead attaches to the silicon wafer contact.

5) Set up the power supply in constant current mode (galvanostat) and make sure that it is set to deliver a constant current density of 40 mA cm^{-2} (corresponding to a current of 48 mA if the Standard etch cell is used).

6) Turn on the illumination lamp, center and focus it on the chip. The illumination source should deliver ~150 mW cm^{-2} to the silicon surface. Light intensity can be measured prior to setting up the etch cell with a light power meter.

7) Activate the power supply and apply the anodic current (40 mA cm^{-2}) for 10 minutes.

8) After completion of the etch, turn off the power source and remove the HF solution from the cell. Rinse the cell with ethanol, remove the chip and rinse it again with ethanol, then blow it dry under a stream of nitrogen.

9) Illuminate the sample with the UV LED in a darkened room. An orange glow should be apparent (Figure 2.6a).

Results:

The luminescence will fade a bit as the sample ages. This comes from an oxide layer that grows on the silicon nanocrystalites as they sit in air. The oxide that grows on porous silicon made from n-type wafers tends to contain a lot of interfacial defects. The defects act as non-radiative recombination centers, trapping electron–hole pairs and causing them to recombine without emission of a photon. The energy of recombination is liberated in the form of heat. In contrast, photoluminescence from porous silicon made from p-type wafers tends to be pretty dim at first, increasing in intensity as the sample ages. Presumably the nanocrystallites in this type of material are not small enough to display quantum confinement effects right out of the etch bath, and as the oxide grows the silicon features shrink into the quantum size regime. This oxide apparently passivates the silicon nanocrystallites.

The effect of aging on luminescence intensity, and the different behavior observed for n- and p-type material, has been noted by many workers. The origin of this difference remains obscure. Presumably it is driven by the different morphologies produced in the etching process; p-type material tends to be predominantly microporous, with fine nanometer-scale silicon filaments, while n-type etches into a mixture of macro-, meso-, and micropores and contains a wide range of silicon feature sizes (Figure 2.6b).

When Canham first prepared luminescent porous silicon he let the p-type sample sit in the electrolyte for a few hours after turning off the power supply. The O_2 present in the air slowly oxidized the silicon nanostructures, and the oxide was simultaneously removed by HF present in the electrolyte. The silicon features shrank into the quantum size regime, and a luminescent sample resulted [1]. A variety of methods can be used to accomplish the thinning process needed to get photoluminescence out of p-type porous silicon. Exposure of a dry sample to air for a few weeks is sufficient, though you can speed the process by thermal oxidation in an oven at 200 °C or higher [10]. In the case of p-type porous silicon, the oxide formed at the surface appears to passivate the surface of the nanostructures without generating additional non-radiative traps. Light will photocorrode the nanostructures during (or after) the etch to generate more of the smallest features needed, and p-type samples illuminated with blue light during etching will come out of the bath displaying reasonably bright photoluminescence [11].

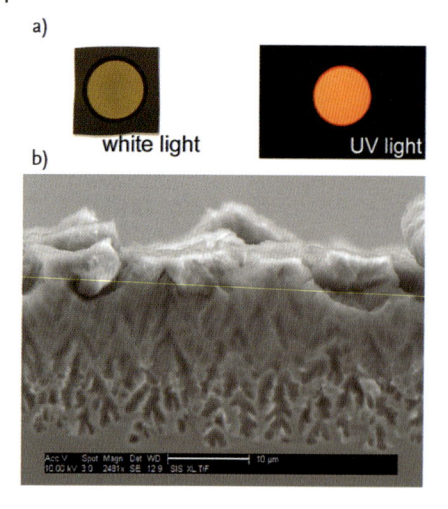

Figure 2.6
(a) Photos of an n-type, macroporous silicon sample, prepared by frontside illumination. The sample has a matted, dull tan appearance under room lights. The orange photoluminescence is apparent under UV (380 nm) illumination. (b) Cross-sectional scanning electron microscope (SEM) image showing the branched, macroporous morphology. Images courtesy Luo Gu, UCSD.

or bubbles in the cell. This next experiment provides an example of how to etch via back side illumination. The standard cell design (Appendix 1) has a hole drilled in the back that allows you to illuminate the back side of the wafer with white light coupled through an optical fiber. A suitably placed near-infrared LED will also work.

Back side etching can be performed in either galvanostatic (constant current) or potentiostatic (constant voltage) modes. The advantage of the potentiostatic mode is that the etch rate can be adjusted by changing the illumination intensity. There is a roughly linear relationship between porosity and light intensity in potentiostatic mode [12]. This is convenient for modulating the pore diameters in the direction of pore propagation in macroporous structures; the intensity of the light source can be modulated programmatically using a microcomputer.

2.11
Preparation of Porous Silicon by Stain Etching

Stain etching is an alternative to electrochemical anodization that is useful if one desires to start from silicon powders or other forms of silicon that cannot conveniently be connected to an electrode. It readily scales to large

EXPERIMENT 2.4: Photoelectrochemical etch of an n-type silicon wafer by back side illumination

In this experiment you will prepare a luminescent porous silicon sample from an n-type wafer by back side illumination. This procedure produces a macroporous sample that is ~60 μm thick, with pore diameters of the order of 2 μm.

Equipment/supplies:

Constant current power supply	See Table 2.1 for examples. For this experiment, a Princeton Applied Research PAR 363 potentiostat/galvanostat was used, operating in galvanostatic mode
Pt mesh counter electrode	See Table 2.1. A loop or coil can also be used
Al foil back-contact	This should be fabricated from a heavy gauge aluminum foil – aluminum weighing boats from Alfa Aesar (www.alfa.com) can be cut to size. A hole must be punched in the center of the electrode to allow illumination of the silicon electrode. A common paper hole punch can be used. Alternatively, a glass slide coated with the transparent conductor indium-doped tin oxide (ITO) can be used (Ted Pella, inc).
Small etch cell	The small etch cell (0.2 cm^2 area) is used for this experiment, Appendix 1. The diameter of the porous silicon film prepared is determined by the illuminated area of the electrode
n-type (100) Si chip, resistivity 0.8 Ω-cm	Siltronix (www.siltronix.com). Measurement of resistivity is performed in Experiment 1.1
1:1 aqueous HF (48%): ethanol solution	See Section 2.5
Light source	Dolan-Jenner MI-150 high intensity fiber illuminator (www.dolan-jenner.com). The fiber must be fed into the hole drilled in the base piece of the etch cell.

Procedure:

1) Rinse the n-type silicon chip with ethanol, dry it, and weigh it.

2) Mount the chip in the small etch cell (Appendix 1); the back-contact should be a piece of aluminum foil containing a punched hole. The hole in the aluminum foil must line up with the hole in the base of the etch cell.

3) Place your cell in the fume hood designated for handling HF and add the 1:1 (v/v) mixture of 48% aqueous HF and absolute ethanol.

4) Immerse the platinum counter-electrode in the electrolyte. Attach the counter-electrode to the negative (black) lead of the power supply. Attach the positive (red) lead to the aluminum foil back-contact. If you are using an electrochemical power supply such at the Princeton Applied Research PAR 363, the reference and counter-electrode leads must be shorted together and connected to the platinum electrode; the working electrode lead attaches to the silicon wafer contact.

5) Set the power supply to constant voltage mode (potentiostat) and prepare it to apply a potential of 1.5 V.

6) Turn on the illumination lamp, illuminate the backside of the wafer, and set the lamp to the appropriate intensity. The illumination source should be bright enough to allow 40 mA cm^{-2} to pass (corresponding to a current of 8.4 mA if using the small etch cell) when the potentiostat is switched on. You may need to determine this with a sacrificial silicon wafer in a separate experiment (the pore dimensions are controlled by light intensity; if you play with light intensity during the etch the porous layer will not be uniform).

7) Activate the power supply and apply the potential for 10 minutes.

8) After completion of the etch, turn off the power source and remove the HF solution from the cell. Rinse the cell with ethanol, remove the chip and rinse it again with ethanol, then blow it dry under a stream of nitrogen.

Results:

The pore dimensions with this type of etch scale with the width of the space charge region in the silicon sample [13, 14]. Thus the pores will be larger for wafers with larger resistivity. The image of Figure 2.7 shows a typical morphology, with well-defined, straight macropores. Samples with pores ordered in the x–y plane of the wafer can also be generated, if the face of the wafer is pre-patterned using a photolithographic process [14].

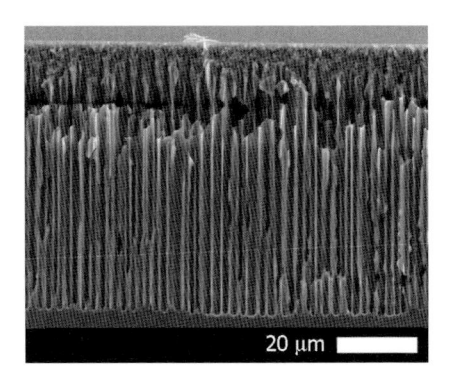

Figure 2.7
Cross-sectional SEM image of n-type porous silicon prepared by back side illumination under potentiostatic control. Image courtesy Joseph Lai, UCSD.

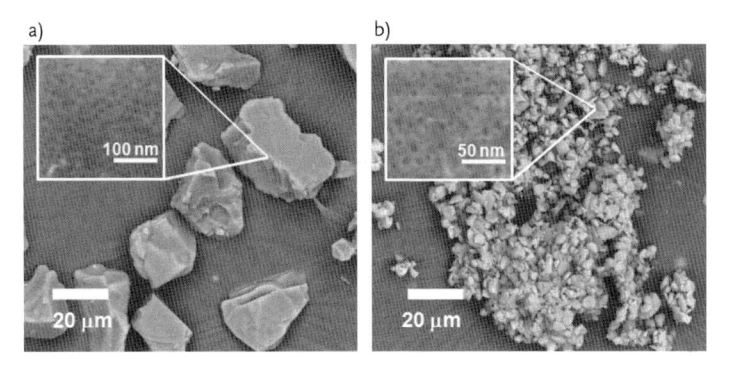

Figure 2.8
SEM images of stain-etched porous silicon powders. Two particle sizes are shown: (a) 20 μm, and (b) 4 μm. Insets show the pore structure. Image courtesy Jennifer S. Park, UCSD.

batches, and it is the main method used to prepare porous silicon powders commercially ("VestaPSi", from Vesta Ceramics, vestaceramics.net). Figure 2.8 shows SEM images of stain-etched porous silicon powders. Because it lacks the programmatic control one obtains with the electrochemical process, stain etching cannot be used to prepare stratified structures such as double layers or multilayered photonic crystals.

The term "stain etching" refers to the brownish or reddish color of the film of porous silicon that is generated on a crystalline silicon wafer subjected to the process [15]. In the stain etching procedure, a chemical oxidant replaces the power supply used to drive the electrochemical reaction. As with electrochemical anodization, HF is a key ingredient (needed to remove silicon oxide), and various other additives are used to control the reaction [16]. Stain etching generally is less reproducible than the electrochemical process, although the reliability of the process has improved considerably compared with the earliest recipes [17, 18]. Nitric acid is the most common oxidant used, although Experiment 2.5 uses vanadium pentoxide due to its better reliability and reproducibility.

EXPERIMENT 2.5: Stain etch of p-type silicon

In this experiment you will prepare a luminescent porous silicon sample from a p-type wafer by stain etching. This procedure produces a mixed micro/mesoporous layer that is ~10 µm thick, with pore diameters of the order of 10 nm.

Equipment/supplies:

50 mL Teflon or polypropylene beaker	A glass beaker cannot be used because of the HF content of the etchant.
p-type (100) Si chip, resistivity 1 Ω cm	Siltronix (www.siltronix.com). Measurement of resistivity is performed in Experiment 1.1
0.05 M V_2O_5 in 1:1 aqueous HF (48%): ethanol solution	Other oxidants, such as Fe(III), MnO_4^- or HNO_3 can be substituted for V_2O_5

Procedure:

1) Clean the p-type silicon chip in isopropanol in an ultrasonic cleaner for 15 minutes.

2) In a fume hood designated for handling of HF and corrosive solutions, place 10 ml of the etchant solution (0.05 M V_2O_5 in 1:1 aqueous 48% HF:ethanol) in a Teflon beaker.

3) Using Teflon tweezers, carefully place the silicon chip in the etchant solution for 3 minutes.

4) Remove the silicon chip from the cell, rinse it with ethanol, then blow it dry under a stream of nitrogen.

Results:

This sample will display relatively strong photoluminescence in the red to red-orange spectral region. The etchant has limited shelf-life, so it is best to make it fresh each time. With minor modifications this procedure can be used to stain etch silicon particles or other unusual shapes. The procedure calls for a Teflon or Nalgene beaker, but you can also use one of the etch cells from Appendix 1. Techniques for measurement of the photoluminescence spectrum are described in Chapter 5 (Section 5.2).

2.12
Preparation of Silicon Nanowire Arrays by Metal-Assisted Etching

The last preparative method of this chapter focuses on a somewhat different and unusual form of porous silicon–nanowire arrays. The procedure utilizes metal-assisted etching, where metallic nanoparticles are deposited on a silicon wafer surface to catalyze the etching process. Discovered by Kelly et al. in the late 1990s [19, 20] and adapted for nanowire production by Peng in 2005 [21], metal-assisted etching is a convenient and inexpensive alternative to the more established means to prepare silicon nanowires such as vapor–liquid–solid (VLS) growth [21–23]. Hydrogen peroxide is the usual oxidizing agent, and the noble metal can be deposited by electroless deposition, photopatterning, or ion implantation [24, 25]. During the reaction, valence band holes are injected into the silicon substrate by means of the decomposition of H_2O_2 on the metal surface, which causes the etching reaction to be localized in the vicinity of the metal. If the metal exists on the surface as a dense array of particles or nanoparticles, pore formation causes the metal particles to drill into the surface of the wafer, leaving behind an array of aligned silicon nanowires. Figure 2.9 shows a cross-sectional SEM image of silicon nanowires produced by metal assisted stain etching of p-type (100) silicon.

This experiment utilizes the electroless silver nanoparticle-assisted etching technique developed by Peng and coworkers [26–28]. The silver nanoparticles are deposited onto the surface of the substrate by electroless deposition from an aqueous solution. The electrolyte for the etching reaction is a mixture of H_2O_2, aqueous HF and ethanol.

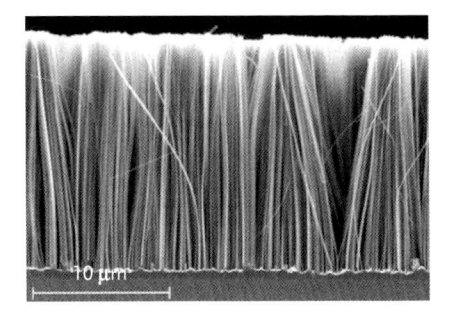

Figure 2.9
Cross-sectional SEM image of silicon nanowires produced by metal-assisted stain etching of p-type (100) silicon. Image from reference [29].

EXPERIMENT 2.6: Synthesis of silicon nanowires by metal-assisted stain etch of p-type silicon

In this experiment you will prepare a silicon nanowire array from a p-type wafer by metal-assisted stain etching. This procedure produces a macroporous sample that is ~15 µm thick, with wire diameters of the order of 50 nm.

Equipment/supplies:

Standard etch cell	The Standard etch cell (1.2 cm² area) is used for this experiment, Appendix 1.
p-type (100) silicon chip, resistivity 10 Ω cm	Siltronix (www.siltronix.com). Measurement of resistivity is performed in Experiment 1.1.
0.2 M $AgNO_3$ in 10% aqueous HF	Used to deposit the silver metal nanoparticles. In a Teflon or Nalgene beaker, dissolve 0.7 g of $AgNO_3$ (Sigma-Aldrich) in a solution containing 5 ml of aqueous 49% HF and 20 ml of deionized water.
Aqueous etchant solution, 10% in HF and 0.6% in H_2O_2	Used to etch the silicon nanowires. In a Teflon or Nalgene beaker, add 5 ml of aqueous 49% HF to 19.5 ml of deionized water. Add 0.5 ml of 30% H_2O_2 (Sigma-Aldrich) to this solution.

Procedure:

1) Clean the p-type silicon chip in isopropanol in an ultrasonic cleaner for 15 minutes.

2) Mount the chip in the Standard etch cell and place it in a fume hood designated for handling of HF and corrosive solutions.

3) Clean the wafer by removal of a sacrificial porous silicon layer (see Section 2.6.4): electrochemically etch the sample in 3:1 aqueous 48% HF:ethanol at 100 mA cm^{-2} for 30 seconds. Remove the electrolyte and add 1 M aqueous NaOH solution containing 5% ethanol for 5 minutes. Rinse the cell with ethanol and dry.

4) Add the aqueous $AgNO_3$/HF solution for 1 minute, rinse with deionized water.

5) Add the 10% HF + 0.6% H_2O_2 etchant solution for 30 minutes. Rinse the cell thoroughly with deionized water.

6) Remove the silicon chip from the cell, rinse it with ethanol and then with pentane, then blow it dry under a gentle stream of nitrogen.

Results:

The material resulting from this procedure will appear black due to efficient light scattering from the pillars. The silver nanoparticle catalyst is not removed in the process; most of it can be found at the base of the pores. The etchant has limited shelf-life, so it is best to make it fresh each time.

References

1. Canham, L.T. (1990) Silicon quantum wire array fabrication by electrochemical and chemical dissolution. *Appl. Phys. Lett.*, **57** (10), 1046–1048.
2. Lehmann, V., and Gosele, U. (1991) Porous silicon formation: a quantum wire effect. *Appl. Phys. Lett.*, **58** (8), 856–858.
3. Zhang, X.G. (2001) *Electrochemistry of Silicon and Its Oxide*, Kluwer Academic–Plenum Publishers, New York.
4. Schwartz, M.P., Derfus, A.M., Alvarez, S.D., Bhatia, S.N., and Sailor, M.J. (2006) The smart petri dish: a nanostructured photonic crystal for real-time monitoring of living cells. *Langmuir*, **22**, 7084–7090.
5. Pacholski, C., Yu, C., Miskelly, G.M., Godin, D., and Sailor, M.J. (2006) Reflective interferometric fourier transform spectroscopy: a self-compensating label-free immunosensor using double-layers of porous SiO2. *J. Am. Chem. Soc.*, **128**, 4250–4252.
6. Pacholski, C., Sartor, M., Sailor, M.J., Cunin, F., and Miskelly, G.M. (2005) Biosensing using porous silicon double-layer interferometers: reflective interferometric Fourier transform spectroscopy. *J. Am. Chem. Soc.*, **127** (33), 11636–11645.
7. Janshoff, A., Dancil, K.-P.S., Steinem, C., Greiner, D.P., Lin, V.S.-Y., Gurtner, C., Motesharei, K., Sailor, M.J., and Ghadiri, M.R. (1998) Macroporous p-type silicon Fabry-Perot layers. Fabrication, characterization, and applications in biosensing. *J. Am. Chem. Soc.*, **120** (46), 12108–12116.
8. Nash, K.J., Calcott, P.D.J., Canham, L.T., and Kane, M.J. (1994) The origin of efficient luminescence in highly porous silicon. *J. Lumin.*, **60–61**, 297–301.
9. Calcott, P.D.J., Nash, K.J., Canham, L.T., Kane, M.J., and Brumhead, D. (1993) Spectroscopic identification of the luminescence mechanism of highly porous Si. *J. Lumin.*, **57**, 257–269.
10. Petrova-Koch, V., Muschik, T., Kux, A., Meyer, B.K., Koch, F., and Lehmann, V. (1992) Rapid thermal oxidized porous Si- the superior photoluminescent Si. *Appl. Phys. Lett.*, **61** (8), 943–945.
11. Doan, V.V., Penner, R.M., and Sailor, M.J. (1993) Enhanced photoemission from short-wavelength photochemically etched porous silicon. *J. Phys. Chem.*, **97**, 4505–4508.
12. Matthias, S., Schilling, J., Nielsch, K., Muller, F., Wehrspohn, R.B., and Gosele, U. (2002) Monodisperse diameter-modulated gold microwires. *Adv. Mater.*, **14** (22), 1618–1621.
13. Foll, H., Christopherson, M., Carstensen, J., and Haase, G. (2002) Formation and application of porous

14 Lehmann, V. (1993) The physics of macropore formation in low doped n-type Si. *J. Electrochem. Soc.*, **140** (10), 2836–2843.

15 Archer, R.J. (1960) Stain films on silicon. *J. Phys. Chem. Solids*, **14**, 104–110.

16 Kelly, M.T., Chun, J.K.M., and Bocarsly, A.B. (1994) High-efficiency chemical etchant for the formation of luminescent porous silicon. *Appl. Phys. Lett.*, **64** (13), 1693–1695.

17 Turner, D.R. (1960) On the mechanism of chemically etching germanium and silicon. *J. Electrochem. Soc.*, **107**, 810.

18 de Vasconcelos, E.A., da Silva, E.F., dos Santos, B., de Azevedo, W.M., and Freire, J.A.K. (2005) A new method for luminescent porous silicon formation: reaction-induced vapor-phase stain etch. *Phys. Status Solidi A-Appl. Mater.*, **202** (8), 1539–1542.

19 Ashruf, C.M.A., French, P.J., Bressers, P.M.M., and Kelly, J.J. (1999) Galvanic porous silicon formation without external contacts. *Sens. Actuators A*, **74**, 118.

20 Xia, X.H., Ashruf, C.M.A., French, P.J., and Kelly, J.J. (2000) Galvanic cell formation in silicon/metal contacts: the effect on silicon surface morphology. *Chem. Mater.*, **12**, 1671.

21 Peng, K.Q., Wu, Y., Fang, H., Zhong, X.Y., Xu, Y., and Zhu, J. (2005) *Angew. Chem. Int. Ed. Engl.*, **44**, 2737.

22 Xia, Y., Yang, P., Sun, Y., Wu, Y., Mayers, B., Gates, B., Yin, Y., Kim, F., and Yan, H. (2003) One-dimensional nanostructures: synthesis, characterization, and applications. *Adv. Mater.*, **15** (5), 353–389.

23 Kolasinski, K.W. (2005) Silicon nanostructures from electroless electrochemical etching. *Curr. Opin. Solid State Mater. Sci.*, **9** (1–2), 73–83.

24 Li, X., and Bohn, P.W. (2000) Metal-assisted chemical etching in HF/H_2O_2 produces porous silicon. *Appl. Phys. Lett.*, **77**, 2572.

25 Harada, Y., Li, X., Bohn, P.W., and Nuzzo, R.G. (2001) Catalytic amplification of the soft lithographic patterning of Si. Nonelectrochemical orthogonal fabrication of photoluminescent porous Si pixel arrays. *J. Am. Chem. Soc.*, **123**, 8709.

26 Peng, K.Q., Zhang, M.L., Lu, A.J., Wong, N.B., Zhang, R.Q., and Lee, S.T. (2007) Ordered silicon nanowire arrays via nanosphere lithography and metal-induced etching. *Appl. Phys. Lett.*, **90** (16), 163123.

27 Peng, K.Q., Fang, H., Hu, J.J., Wu, Y., Zhu, J., Yan, Y.J., and Lee, S. (2006) Metal-particle-induced, highly localized site-specific etching of Si and formation of single-crystalline Si nanowires in aqueous fluoride solution. *Chem. Eur. J.*, **12** (30), 7942–7947.

28 Peng, K.Q., Wu, Y., Fang, H., Zhong, X.Y., Xu, Y., and Zhu, J. (2005) Uniform, axial-orientation alignment of one-dimensional single-crystal silicon nanostructure arrays. *Angew. Chem. Int. Ed.*, **44** (18), 2737–2742.

29 Peng, K., Lu, A., Zhang, R., and Lee, S.-T. (2008) Motility of metal nanoparticles in silicon and induced anisotropic silicon etching. *Adv. Funct. Mater.*, **18**, 3026–3035.

3
Preparation of Spatially Modulated Porous Silicon Layers

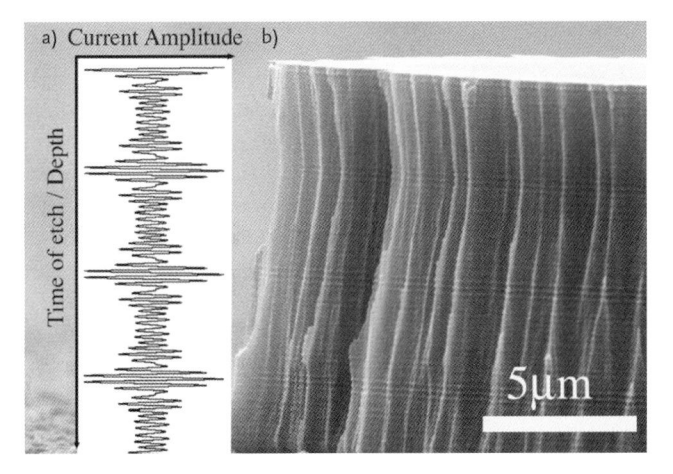

Figure 3.1
Cross-sectional SEM image showing the mapping of a current–time waveform to a porosity gradient in porous silicon. The waveform (a) contains five temporal frequencies. The image (b) shows the direct mapping of the temporal frequencies to the spatial frequencies of the porosity modulation. From reference [1].

This chapter describes etching of more complicated structures from porous silicon, in which the pore diameters or the porosity is modulated in one or more directions. Here, we will describe preparation of gradients, double layers, photonic crystals, and surfaces patterned in the x–y plane. Porosity modulation in the z-direction is accomplished by changing the applied current during the etch, while modulation in the x–y plane is accomplished by lithographic masking or photopatterning. The exercises in this chapter leave out some of the details that were included in the experiments of Chapter 2. Refer to those experimental descriptions if you have trouble with any of the procedures described in this chapter. The coupling of chemical

Porous Silicon in Practice: Preparation, Characterization and Applications, First Edition.
Michael J. Sailor.
© 2012 Wiley-VCH Verlag GmbH & Co. KGaA. Published 2012 by Wiley-VCH Verlag GmbH & Co. KGaA.

modification of porous silicon with spatially modulated etches is treated separately in Chapter 6.

3.1
Time-Programmable Current Source

The most efficient means to prepare most of the structures described in this chapter is with a computer-controlled power supply. Since the porosity and pore size at any instantaneous point in time during an etch are a direct map of the current density, if you control the current–time relationship, you control the porosity–depth relationship in the film. There are commercial devices available that can be programmed to source a current for a prescribed period of time, then step the current to a different value for an additional time period. The Keithley Sourcemeter series 2100 is a good example. While this is adequate to prepare double layers or Bragg stacks, more complicated pore structures require greater programmatic control of the etching waveform. For these structures you will need to interface a microcomputer with the power supply.

There are two paths that can be followed to implement computer control of the etching source, digital or analog, and these are somewhat dependent on personal preference. In either case, the waveform is generated in software, using National Instruments Labview® (www.ni.com), Mathworks Matlab® (www.mathworks.com), Microsoft Excel® (www.microsoft.com), Wavemetrics IGOR Pro® (www.wavemetrics.com) or the like. The waveform is then transmitted to the etching power supply as either an analog or a digital signal. If analog, the computer must be equipped with a digital to analog converter (DAC) card and the power supply must be able to convert an analog voltage input to a current output. If digital, the power supply must have the appropriate interface and onboard DAC. The main trade-off is speed. For rugate filters and more complicated structures, the data rate should be at least 500 points s^{-1} (see time resolution discussion below). Most DAC cards can easily output an analog signal at this data rate. In principle the rate is also sufficient for a digital interface (GPIB, USB, etc), but all systems on the market today (of which the author is aware) do their hardware/software handshaking between the power supply and the computer using a point-by-point, rather than a buffered data transfer protocol. As of this writing, the power supply manufacturers investigated had not implemented software drivers that allow buffered data transfers. The 500 points s^{-1} limitation of the digital systems available today is not a serious limitation (see below), and the simplest solution is an all-digital one. The author prefers the Keithley Sourcemeter® family coupled to a microcomputer running National Instruments Labview® software and

Table 3.1 Time response of etch cells used in current-programmed etches.

Etch cell area (cm²)[a]	95% response time (ms)[b]	Waveform maximum frequency (Hz)[c]	System data rate (points s⁻¹)[d]
1.2 (Standard etch cell)	7.2	140	500
0.21 (Small etch cell)	1.3	790	2500
8.6 (Large etch cell)	52	19	60

a) Etch cell sizes refer to the cell designs in Appendix 1.
b) Time = 3τ (Equation 3.1). Assumes $R_s = 100\,\Omega$, $C_d = 20\,\mu F\,cm^{-2}$.
c) The maximum data rate for the waveform. If the frequency of the waveform exceeds this value, the current response of the cell will not follow the waveform and some smoothing will occur.
d) Recommended values, corresponding to the sample rate of the DAC card or the digital power supply.

PCIB card (see below). Software to generate and etch waveforms for these systems is available on the author's website (http://sailorgroup.ucsd.edu/software/).

If you do not mind programming, the analog solution is more versatile: generate the current–time waveform in a computer program, output the analog waveform with a DAC card, feed it into the analog input of a voltage-programmable current source, such as the Kepco operational amplifier-based power ATE supply or a PAR 263 (Table 3.1). This is also the less expensive route. Keep in mind that the term "programmable power supply" used to describe many low-cost systems generally refers to a device with which you can set a constant current (or voltage) via front panel switches or a computer interface; it does not mean that you will be able to dither the current on the time scales needed to prepare most of the structures in this chapter. The proper term for the analog waveform etching system needed here is "voltage programmable current source".

3.1.1
Time Resolution Issues

The time resolution of the etching system in most cases is set by the series resistance of the cell and the capacitance of the silicon/electrolyte interface. The capacitance at an aqueous electrolyte interface is determined by the electrical double layer in the solution, which is potential dependent. A value of $20\,\mu F\,cm^{-2}$ is typical. For a semiconductor electrode, the space charge region adds additional (series element) capacitance. For low-doped semiconductors, space charge capacitance can be substantial, while for highly doped material it is negligible. Considering only the solution (double layer) contribution to capacitance, the minimum time constant τ for the system can be defined as

$$\tau = R_s C_d \tag{3.1}$$

where R_s is the total series resistance and C_d is the double layer capacitance. The value of τ represents the exponential decay constant of the electrode. After the power supply imposes a change in current, it takes 3 time constants for the current at the cell to reach 95% of the set value. These response times and the maximum frequencies of the corresponding waveforms that can be etched are given in Table 3.1 for the three types of etch cells used in this book (Appendix 1). Assuming series resistance (100 Ω) and double layer capacitance (20 $\mu F\,cm^{-2}$) values typical of an aqueous HF/ethanol electrolyte, the response time is of the order of 10 ms for the Standard etch cell (Appendix 1). Thus it makes no sense to etch a waveform with frequency components in excess of 150 Hz. The small etch cell, with its smaller exposed area of 0.21 cm^2 (Appendix 1), can handle somewhat higher frequencies of ~1 kHz. Frequencies in excess of 1 kHz are generally not needed for etching photonic structures in porous silicon, as the thickness of porous silicon generated in 1 ms of etching at 200 mA cm^{-2} (a typical current density) is less than 1 nm. The spatial resolution of the etch cannot exceed the average pore size.

Somewhat related to the frequency of the waveform is the data rate that can be output by the etching system. The data rate is defined as the number of unique current values, or points, output per second. The Nyquist–Shannon theorem tells us that the number of data points (the sampling frequency) must be greater than twice the waveform bandwidth in order to adequately represent the waveform. When the DAC subsystem outputs too few points per second to reproduce the highest frequencies in the waveform, it is said to be undersampling. To avoid undersampling, the data rate of the system should be greater than twice the maximum frequency of the waveform that you wish to etch. The data rate of an etching system is determined by either the DAC card, the digital power supply, or in some cases, the computer program; recommended values ("System data rate") are given in Table 3.1.

3.1.2
Etching with an Analog Source

The basic components of a computer-controlled etching rig that accepts an analog waveform are presented in Table 3.2. It consists of a personal computer equipped with a DAC card, an interface board, and a voltage to current converter. Many DAC/digital computer combinations can send out a time-varying waveform at >2000 points s^{-1}, but few voltage to current converters can slew the current that fast. The minimum frequency response of the

Table 3.2 Power supply used for current-programmed etches of porous silicon (digital to analog solution).

Item	Vendor	Comments
Programmable current/voltage source: ATE 25-2DM	Kepco (www.kepco.com)	Great etching tool based on an op-amp controlled current source. Low noise, linear response, can deliver up to 2 A. Digital control by GPIB interface (via National Instruments Labview program) too slow to handle high data rates needed to prepare rugate or other photonic crystal structures as of this writing (a driver to handle buffered i/o had not been implemented). However, the well-behaved analog input (specify the analog input option when ordering) is fast enough (<1 ms) to follow the output from a computer-controlled DAC card, providing great versatility for many user-defined experiments. A bipolar version is also available as the BOP-50-4D (allows you to swing positive and negative of zero, not necessary for etching). An alternative is Princeton Instruments' PAR 263A/94 (see Table 2.2)
Digital to analog converter (plug-in card for a PC): NI PCI-6221 D/A card, 16 Analog Inputs, 24 Digital I/O, 2 Analog Outputs, part # 779066-01	National Instruments (www.ni.com)	You also have to purchase National Instruments' LabView program, an object-oriented programming language that drives the D/A card. A bare-bones etching program can be found at http://sailorgroup.ucsd.edu/software/
Cables and connectors: BNC-2110 Noise Rejecting, Shielded BNC Connector Block, part #777643-01 SHC68-68-EPM Shielded Cable, 68 D-Type to 68 VHDCI Offset, 1 m, part #777643-01	National Instruments (www.ni.com)	
Microcomputer	www.dell.com	Almost any computer that has a slot to accept a PCI card will work. Dell dimension 9200 or equivalent

power supply should be 0.5 kHz. The author's research group has had much success with the Kepco ATE series.

3.1.3
Etching with a Digital Source

The components of a computer-controlled digital etching rig are presented in Table 3.3. While the digital interface card and computer can send out a time-varying waveform at >1.5 MB s^{-1}, I/O overhead and other limitations of many all-digital instruments set the current waveform output rate to 500 points s^{-1}. However, the time resolution of the system in most cases is set by the series resistance of the cell, and the capacitance of the silicon/electrolyte interface, and this is adequate for all the experiments described in this book.

Table 3.3 Power supply used for current-programmed etches of porous silicon (all digital solution).

Item	Vendor	Comments
Programmable current/voltage source: 2601 Sourcemeter	Keithley (www.keithley.com)	Can program simple waveforms from front panel, or more complicated ones by computer interface. As of this writing the interface handles data rates up to 500 points s^{-1}, just adequate to prepare rugate or other photonic crystal structures – the version we tested produces short current spikes when changing scales or settings so it should be used with the automatic current range adjustment disabled.
Digital interface card (GPIB plug-in card for a PC): PCI-GPIB, NI-488.2 for Windows 2000/XP, 2M X2 Cable, Part #778032-51	National Instruments (www.ni.com)	To allow computer interface.
Microcomputer	www.dell.com	Almost any computer that has a slot to accept a PCI card will work. Dell dimension 9200 or equivalent

3.2
Pore Modulation in the z-Direction: Double Layer

To prepare this sample (Experiment 3.1), you only need to step the current from a higher to a lower value midway through the etch. This can be accomplished manually but a programmed power supply provides better reproducibility. The waveform used and the resulting pore structure it generates are shown in Figure 3.2. This sample possesses a layer of larger pores on top of a layer of smaller pores. The reverse (small pores on top of large pores) can also be prepared. Such structures can be used as size exclusion matrices, to separate biomolecules, for example [2–4]. Although the general structure type is easy to prepare, the nominal pore sizes are highly dependent on dopant concentration and current density; you should verify the pore structure in your sample by electron microscopy or another method if pore diameter is critical to your application.

3.3
Pore Modulation in the z-Direction: Rugate Filter

In Experiment 3.2 we prepare an example of a multilayer, or one-dimensional photonic crystal. The term photonic crystal refers to the fact that the structure interacts with visible or near-infrared light ("photonic") by diffraction from some sort of repeating structure ("crystal") in the material [8]. Many examples of photonic crystals can be found in Nature – opals, butterfly wings, beetle cuticles, and the inside of the abalone shell are familiar examples of what the biologists refer to as "structural color" [9]. In all these structures, the color of the photonic crystal is determined by the spacing and the refractive index of the layers. The layers are evenly spaced, and the refractive index of the layers alternates between two discrete values.

The construction of a photonic crystal by electrochemical etching of silicon was first described by Vincent in 1994 [10]. The key to the discovery was illustrated by Experiment 3.1 in which we saw that porosity, or the amount of silicon dissolved at the pore/silicon interface at any point in time during the etch, directly maps to the instantaneous current being passed. Thus, if the current density is cycled during the etch, the resulting porous film displays a corresponding modulation in porosity with depth. The variation also represents a modulation in refractive index, and that is what allows us to so easily construct photonic structures with porous silicon. The alternating optical density of the repeating layers diffracts and refracts light to produce high reflectivity at predetermined wavelengths (Figure 3.3).

The photonic crystal first prepared by Vincent was a Bragg stack. This is a structure in which the layers are well-defined, with an abrupt change in

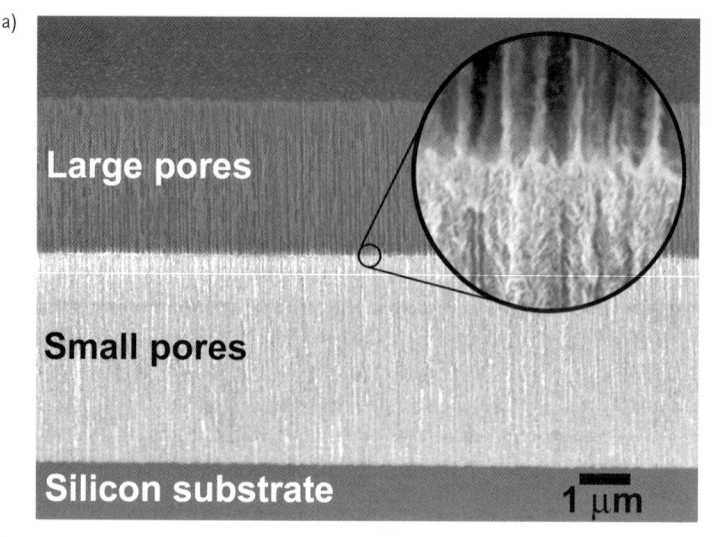

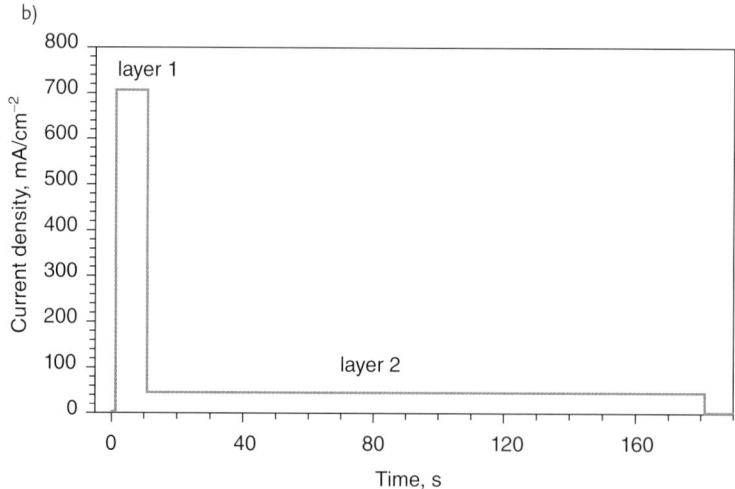

Figure 3.2
Porous silicon double-layer prepared from highly doped p^{++}-silicon (1 mΩ cm). (a) A cross-sectional scanning electron microscope (secondary electron) image of the double layer; (b) the current density versus time waveform used to generate the structure (10 s at 707 mA cm^{-2} followed by 170 s at 44 mA cm^{-2}). After reference [5].

porosity at the interface of each layer. The structure we prepared in Experiment 3.1 can be thought of as a single unit cell of a Bragg stack. In this experiment we will prepare a slightly more complicated photonic crystal, known as a rugate filter. The reflectance spectrum of a rugate filter displays a strong peak (or "stop band"). The term "rugate" refers to the way the refractive index varies in the structure. Unlike the abrupt change between the two refractive index values of a Bragg stack, the refractive index in a

Figure 3.3

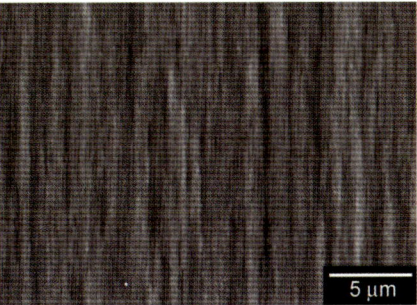

The rugate filter is an example of a 1-dimensional photonic crystal that can be made by etching porous silicon with a time-varying current. A rugate filter consists of a sinusoidal modulation in refractive index, corresponding in this case to a sinusoidal porosity gradient (evident as the horizontal banding in the cross-sectional SEM image). The porous silicon rugate filter reflects a single color, whose wavelength is determined by the periodicity and the refractive index of the layers. SEM image courtesy of Melanie Oakes and Manuel Orosco.

EXPERIMENT 3.1: Mesoporous silicon double layer by programmed electrochemical etch of a p^{++}-type silicon wafer

---

In this experiment you will prepare a double layer of porous silicon from a highly doped p^{++}-type wafer. This procedure produces a mesoporous sample that is ~9 μm thick, with nominal pore diameters of the order of 100 nm in the top layer and 10 nm in the lower layer.

Equipment/supplies:

Programmable current power supply	See Table 3.2 for examples. For this experiment, a Kepco ATE 25-2DM power supply was used, operating in voltage-programmed current mode with an analog voltage input from a National Instruments NI PCI-6221 D/A card and a microcomputer running National Instruments' Labview® software. The Labview program subroutines are available at http://sailorgroup.ucsd.edu/software/
Pt spiral counter-electrode	See Table 2.2. A loop or mesh can also be used
Al foil back-contact	This should be fabricated from heavy gauge aluminum foil – aluminum weighing boats from Alfa Aesar (www.alfa.com) can be cut to size

Standard etch cell	Appendix 1
p^{++}-type boron-doped (100) Si chip, resistivity 1 mΩ cm	Siltronix (www.siltronix.com). Measurement of resistivity is performed in Experiment 1.1
3 : 1 aqueous HF (48%) : ethanol solution	See Section 2.5

Procedure:

1) Clean the silicon chip for 15 minutes in an ultrasonic bath in isopropanol, dry it. A more thorough cleaning by etching a sacrificial porous layer (Experiment 2.2) is advised.

2) Mount the chip in the Standard etch cell (Appendix 1) with an aluminum foil back-contact.

3) Add the electrolyte 3 : 1 (v/v) of 48% aqueous HF and absolute ethanol.

4) Immerse the platinum counter-electrode in the electrolyte. Attach the counter-electrode to the negative (black) lead of the power supply. Attach the positive (red) lead to the aluminum foil back-contact.

5) Activate the power supply and run the current step program defined in Figure 3.2. If using the Standard etch cell, the two current density values of 707 and 44 mA cm^{-2} correspond to 848 and 52.8 mA, respectively.

6) After completion of the etch, turn off the power source and remove the HF solution from the cell. Rinse the cell with ethanol, remove the chip and rinse it again with ethanol, rinse it with hexane, then blow it dry under a stream of nitrogen.

Results:

The final rinse with hexane is to minimize the possibility of cracking of the film due to the strong capillary forces and thermal stresses exerted when ethanol evaporates from the pores. Hexane has both a lower surface tension and a lower heat of vaporization than ethanol. The characteristics of the porous layers are summarized in Table 3.4. Note that the results are approximate; there is a strong dependence on the exact resistivity of the wafer used for this type of etch.

Note that the porosity and pore size of the first layer did not change significantly when the second layer was etched. For example, if we had stopped the etch after the "top" layer was etched, the porosity, thickness, and pore diameters of this layer would be the same as it is in the top layer of the double layer [3]. This is a key feature of the electrochemical etching process for porous silicon: once you etch a layer, its morphology is more or less "frozen". This is because silicon is removed almost exclusively at the interface between the porous layer and the silicon substrate. Differences in electrical conductance of crystalline versus porous silicon, depletion of valence band holes in the porous layer, quantum confinement effects, and enhancement of the electric field in the vicinity of the pore tips are all factors considered to play a role in this anisotropy, and it is unique to silicon, its alloys with germanium, and several other III–V type semiconductors [6]. Such a situation does not exist, for example, in the porous alumina system. Porous alumina is prepared by electrochemical anodization of aluminum metal in mineral acids at fairly large applied potentials [7]. Pore formation is a dynamic process in the case of porous alumina, with structural reordering occurring throughout the duration of the etch. Unlike porous silicon, the pore dimensions, orientation, and alignment are constantly changing while a porous alumina sample is prepared.

Table 3.4 Some characteristics of the porous double layer prepared in Experiment 3.1.

Layer	Porosity (%)	Thickness (nm)	Pore diameters (nm)
Top	76	3200	100
Bottom	38	4400	6

rugate filter varies in a smooth, sinusoidal fashion. A rugate is sometimes preferred over the more common Bragg stack because its spectrum does not display very pronounced higher order harmonics. Additionally, the fabrication of a Bragg stack (Experiment 3.3) is a bit more involved because the individual layers must be designed with the appropriate combination of porosity and thickness such that they are phase matched ($n_1 L_1 = n_2 L_2$, where n_i and L_i are the refractive index and the thickness of an individual layer i, respectively). The first porous silicon version of a rugate filter was prepared by Berger in 1997 [11].

To prepare a rugate filter, we will deliver a sinusoidal current waveform to the silicon wafer. The relationship between current density and refractive index is only approximately proportional, and so to obtain a true sinusoidal refractive index modulation a somewhat modified sinusoid waveform must be used [12]. However, the deviation is not critical, and a simple sine wave produces a relatively well-behaved peak in the reflectivity spectrum. A portion of the waveform to be used is shown in Figure 3.4. Codes to generate such

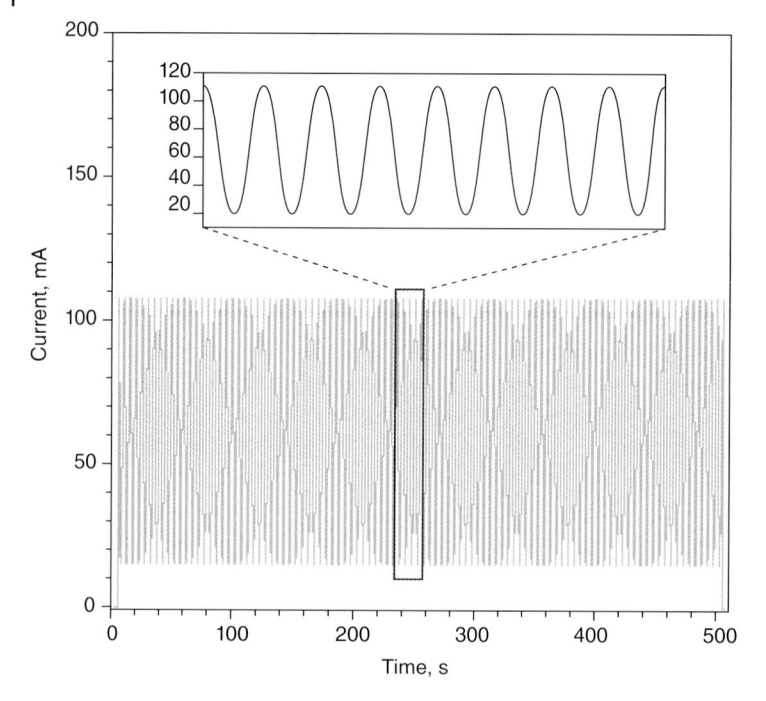

Figure 3.4
Current versus time waveform used to prepare a porous silicon rugate filter from highly doped p^{++}-silicon (1 mΩ cm). The image shows a cosine wave that oscillates between 15 and 108 mA (corresponding to current densities of 12.5 and 90 mA cm^{-2} with the Standard etch cell, Appendix 1). The waveform has 100 cycles, with a period of 6 s, data rate 500 points s^{-1}.

waveforms, written for the National Instruments Labview program, are available at http://sailorgroup.ucsd.edu/software.

3.3.1
Tunability of the Rugate Spectral Peak Wavelength

Equation 3.2 has interesting implications for the design of porous silicon-based rugate filters. In particular, it implies that the spectral position of the reflectivity peak is determined by the average current and the frequency of the sine wave used to etch the sample. These two variables can be controlled independently. Figure 3.6 shows the effect of changing the average current density on the position of the spectral peak. The average current density sets the average porosity and thus the average refractive index (n_{ave}) of the final structure. The relationship of current density to refractive index follows a power law that can be approximated as linear for most p$^+$ and p^{++}

3.3 Pore Modulation in the z-Direction: Rugate Filter | 89

EXPERIMENT 3.2: Multialyered 1-D photonic crystal ("rugate filter") by programmed electrochemical etch of a p^{++}-type silicon wafer

In this experiment you will prepare a 1-D photonic crystal, known as a rugate filter, from a highly doped p^{++}-type silicon wafer. This procedure produces a mixed microporous/mesoporous sample that is 15 μm thick, with nominal pore diameters of the order of 10 nm and average porosity of 60%. The spectral resonance appears at approximately 520 nm

Equipment/supplies:

Programmable current power supply	See Table 3.2 for examples. For this experiment, a Kepco ATE 25-2DM power supply was used, operating in voltage-programmed current mode with an analog voltage input from a National Instruments NI PCI-6221 D/A card and a microcomputer running National Instruments' Labview software. The Labview program subroutines are available at http://sailorgroup.ucsd.edu/software/
Pt spiral counter-electrode	See Table 2.2. A loop or mesh can also be used
Al foil back-contact	This should be fabricated from heavy gauge aluminum foil – aluminum weighing boats from Alfa Aesar (www.alfa.com) can be cut to size
Standard etch cell	Appendix 1
p^{++}-type boron-doped (100) Si chip, resistivity 0.8 mΩ cm	Siltronix (www.siltronix.com). Measurement of resistivity is performed in Experiment 1.1
3:1 aqueous HF (48%): ethanol solution	See Section 2.5

Procedure:

1) Clean the silicon chip for 15 minutes in an ultrasonic bath in isopropanol, then dry it. A more thorough cleaning by etching a sacrificial porous layer (Experiment 2.2) is recommended.

2) Mount the chip in the Standard etch cell (Appendix 1) with an aluminum foil back-contact.

3) Add electrolyte consisting of 3:1 (v/v) mixture of 48% aqueous HF and absolute ethanol.

4) Immerse the platinum spiral counter-electrode in the electrolyte. Attach the counter-electrode to the negative (black) lead of the power supply. Attach the positive (red) lead to the aluminum foil back-contact.

5) Set the power supply to current control mode (voltage limit turned all the way up) with the voltage output from the computer DAC card to the "external" voltage input on the power supply. Note the power supply will be operating in a "voltage programmable current" mode; the machine must have a voltage to current conversion circuit built in. You should test that the computer is controlling the current properly using the test configuration shown in Figure 2.2.

6) Activate the power supply and run the sinusoidal current waveform defined in Figure 3.4 (cosine wave, 6 s period, 100 repeats, minimum and maximum current densities 12.5 and 90.0 mA cm^{-2}). If using the Standard etch cell, the two limiting current density values of 12.5 and 90.0 mA cm^{-2} correspond to 15 and 108 mA, respectively. Note the uniformity of the film will be a little better if you stir or circulate the electrolyte. Stirring prevents hydrogen bubbles from accumulating on the electrodes and ensures a more uniform concentration of HF at the porous silicon surface.

7) After completion of the etch, turn off the power source and remove the HF solution from the cell. Rinse the cell with ethanol, remove the chip and rinse it again with ethanol, then blow it dry under a stream of nitrogen.

Results:

The film is approximately 19 μm thick. The porosity modulation is between 50 and 67%, with an average porosity of 62%. The reflectivity spectrum of a representative sample is shown in Figure 3.5. Chapter 5 provides a detailed description of the procedures to follow and equipment needed to perform the optical measurement. Note the sharp resonance appearing at 520 nm. The wavelength at which this peak occurs is determined by [13]:

$$\lambda_R = 2n_{ave}d_P \qquad (3.2)$$

Where λ_R is the wavelength of the peak, n_{ave} is the average refractive index of the porous silicon layer, and d_P is the spatial periodicity of the nanostructure. The value of d_P for this sample is 190 nm, and the average refractive index of the layer (including the air in the voids) is 1.376.

Using a cosine wave instead of a sine wave (i.e., initial current density at time = 0 of 90.0 mA cm^{-2}) will reduce the intensity of the Fabry–Perot fringes in the spectrum somewhat, because it reduces the reflectivity of the top of the porous film. Significant improvements in the quality of the spectrum can be achieved if the etch begins and ends with current ramps of ~5 s duration. These ramps serve to produce a gradual change in porosity and, thus, refractive index at the air/porous silicon and the porous silicon/silicon substrate interfaces. These gradual index transitions serve as anti-reflection coatings that reduce the intensity of Fabry–Perot fringes in the spectrum. Starting the etch at a current density close to the electropolishing limit (~800 mA cm^{-2}) and then ramping it down to the starting value of the sinusoid, and finishing the etch with a ramp to 0 mA cm^{-2} will produce the desired effect. Apodization of the sinusoid (multiplication by a sin^2 windowing function that tapers the beginning and ending portions of the waveform) can further improve the optical quality of the sample.

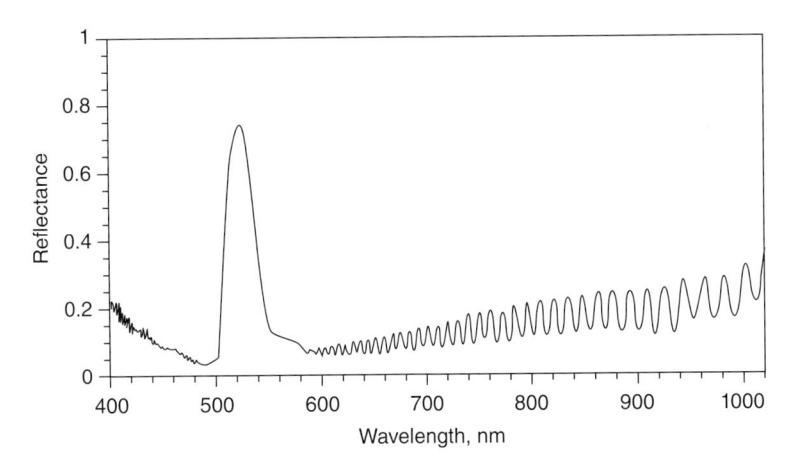

Figure 3.5
Reflectance spectrum of a porous silicon rugate filter prepared using the etching parameters given in Experiment 3.2. Note that the results reported here are approximate; the spectrum will display a moderate dependence on the resistivity of the wafer used. The oscillations in the spectrum from ~580–1000 nm are Fabry–Perot fringes that correspond to reflection from the front and backside of the film (at the air/porous silicon and porous silicon/silicon substrate interfaces).

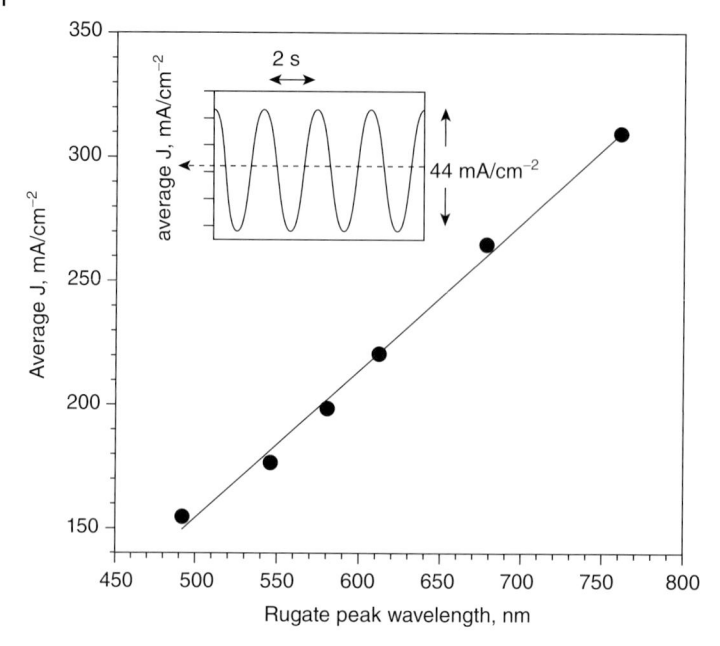

Figure 3.6
Effect of average current density (J) on the position of the main resonance (wavelength of maximum intensity) of a rugate filter. Samples were prepared by etching a sinusoidal waveform (125 cycles, 2 s period) into a 1 mΩ cm p-type wafer, using 3:1 aqueous HF:ethanol. The peak red shifts ~2 nm for every 1 mA cm^{-2} increase in average current density. The amplitude of the current density waveform was held constant at 44 mA cm^{-2}, as shown in the inset.

samples, although the slope of this line depends on dopant density and HF concentration. An example of the relationship of porosity and index to current density for one particular sample/etch type is given in Figure 3.7.

Since the thickness of a given porous layer (d_p) is proportional to the product of current density and time, the wavelength maximum of the peak in the reflectivity spectrum of a porous silicon rugate filter will be determined by the period of the waveform used to prepare it. There is generally a linear correlation of the spectral peak position to period for p$^+$ and p^{++} samples. An example for a p^{++} wafer is shown in Figure 3.8.

3.3.2
Width of the Spectral Band

The width of the spectral band of a rugate filter depends on the number of repeats (cycles) in the waveform and on the magnitude of the difference between the maximum and the minimum refractive index generated by the

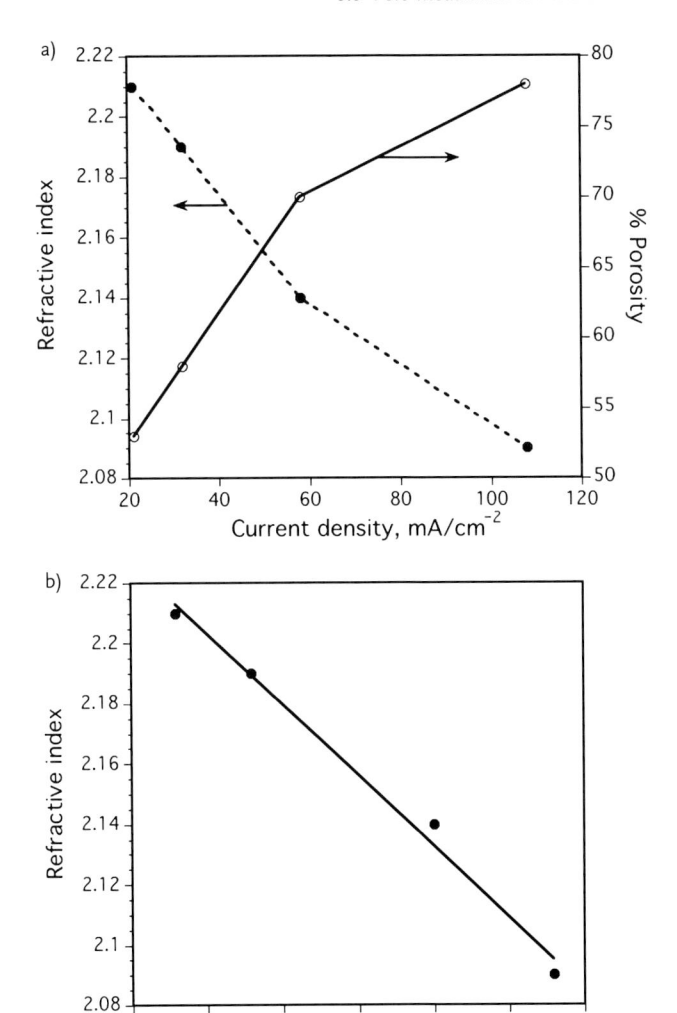

Figure 3.7
Empirical relationship between refractive index and porosity as a function of etch current density for 10 mΩ cm p-type samples, etched in 3:1 aqueous HF:ethanol. (a) Refractive index and porosity as a function of current density. (b) Relationship between refractive index and porosity. Data from ref [12].

sinusoidal waveform. This latter factor is referred to more generally as the index contrast between the layers. The width of a peak narrows with increasing number of repeats and with decreasing index contrast. The relationship between index contrast and peak width is given by Equation 3.3.

$$\frac{\Delta\lambda}{\lambda} \propto \ln\frac{n_{max}}{n_{min}} \tag{3.3}$$

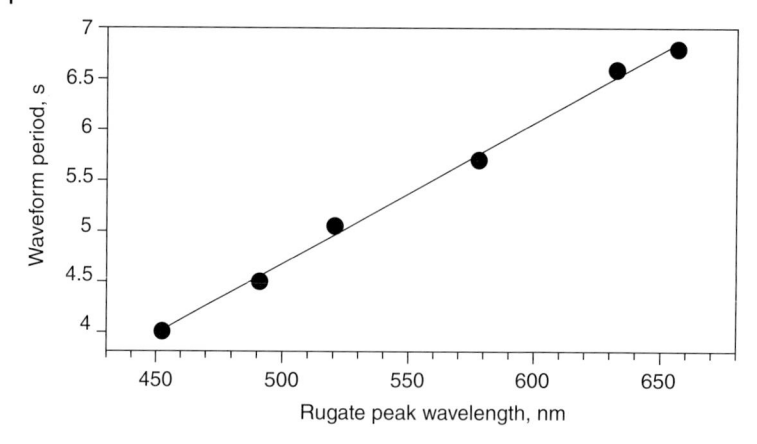

Figure 3.8

Effect of waveform period on the position of the main resonance of a rugate filter. A sinusoidal waveform was used, with a total of 100 cycles and current density maximum and minimum of 108 and 15 mA cm^{-2}, respectively. Samples were prepared from a 0.8 mΩ cm wafer, using 3:1 aqueous HF:ethanol. The peak red shifts ~1 nm for every 14 ms increase in period.

Where n_{max} and n_{min} are the maximum and the minimum refractive index in a cycle, respectively. The linewidth will ultimately be limited by the accuracy with which the refractive index profile matches to a sine wave. The current–porosity relationship changes as the porous film becomes thicker and mass transport into the deep, narrow pores becomes less efficient. The already etched porous layers are also not perfectly static, and the upper layers become more porous as slow chemical dissolution occurs. These factors can be compensated for if very high quality optical structures are needed [12, 14, 15].

3.4
More Complicated Photonic Devices: Bragg Stacks, Microcavities, and Multi-Line Spectral Filters

The purpose of this section is to provide summary procedures for a few of the more complicated optical structures that can be prepared if the appropriate waveform is used in the above experiment. Table 3.5 provides a summary of the device structures and representative spectra. The preferred wafer type for these optical structures is highly doped p-type silicon because it generally provides pore dimensions sufficiently small (<100 nm) that losses due to light scattering are minimized, and it has conductivity sufficient to allow good control of the current. With all of these structures some trial and error is involved in achieving the desired spectrum. This is because

Table 3.5 Some examples of porosity modulated nanostructures that can be prepared with porous silicon. The idealized structures, approximate current density waveforms, and representative spectra are given.[a]

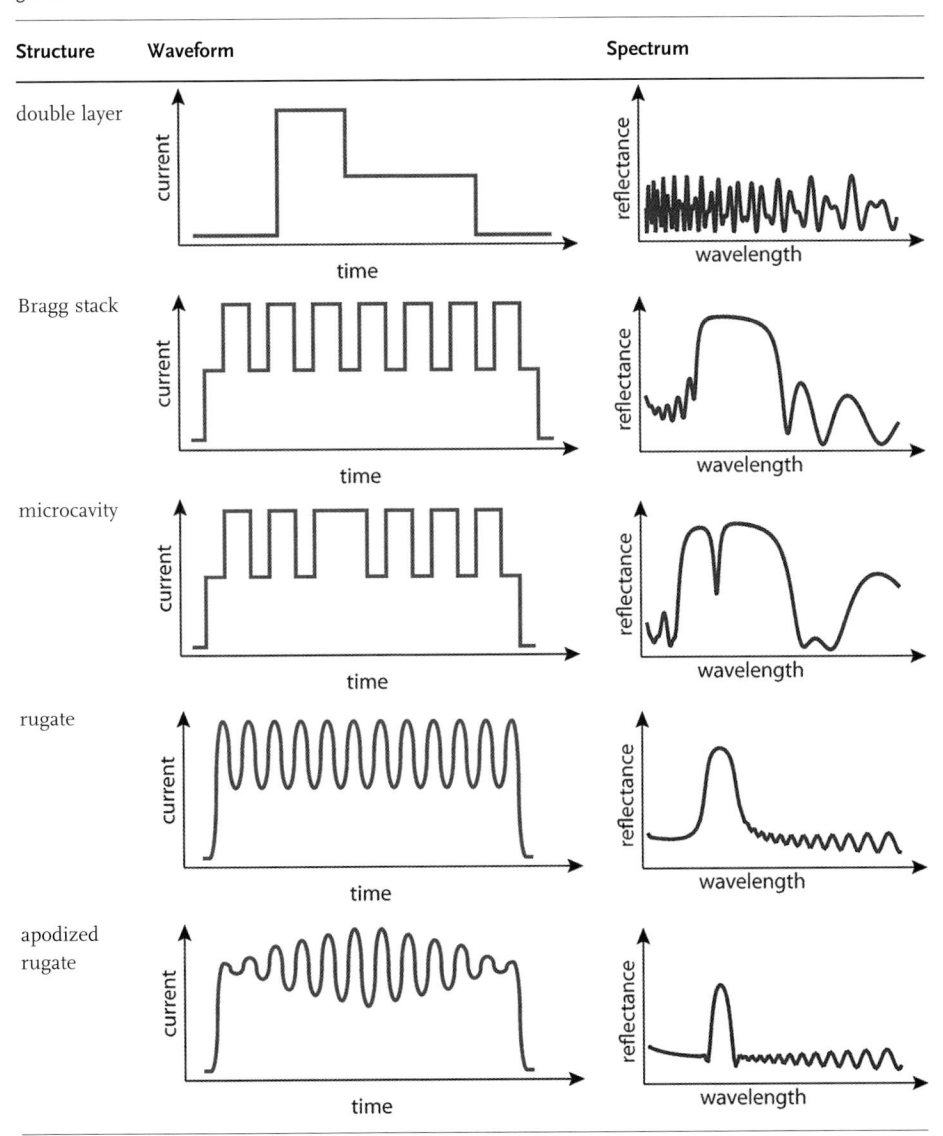

Structure	Waveform	Spectrum
double layer		
Bragg stack		
microcavity		
rugate		
apodized rugate		

a) Spectra calculated using transfer matrix method or interference equation (for double layer). The following parameters were used: (1) double layer, top layer: 3200 nm thick, 76% porosity; bottom layer: 4400 nm thick, 38% porosity. (2) Bragg stack, layer a: 50 nm thick, 43% porosity; layer b: 138 nm thick, 77% porosity, 7 repeats. (3) microcavity, layer a: 64 nm thick, 47% porosity; layer b: 110 nm thick, 81% porosity, 7 repeats. After 3 repeats, a layer 220 nm thick, 81% porous is etched. (4) rugate filter, 30 repeats, porosity minimum = 45%, porosity maximum = 65%. (5) apodized rugate filter, same as (4) but the waveform was apodized with a Gaussian window function prior to etching. Calculated spectra courtesy of Beniamino Sciacca.

porosity is highly dependent on wafer resistivity, temperature, and HF concentration. It is a good idea to carefully measure the resistivity of the wafer you are using (Experiment 1.1) so that you can reproduce the sample on a new wafer. Preliminary experiments in which you etch two single layer samples in order to identify the current–porosity–depth relationship for your experimental set-up can also reduce the amount of trial and error.

3.4.1
Bragg Reflector

Unlike the gradual, continuous modulation of refractive index of a rugate, a Bragg reflector consists of a stack of discrete layers of two different refractive indices. A Bragg stack can display a broader spectral peak than a rugate, providing a true photonic bandgap–rejection of a wide range of wavelengths of light. This band of highly reflected light is called the stop band of the structure. To etch a Bragg properly, each layer has to be 1/4 of the optical thickness (nL) of the design wavelength, and the layers have to be phase matched. This means that you have to keep the product nL constant between layer a and layer b of the ab stack. However, the refractive index (n) of layer a must be different from layer b. That is, the index contrast (Δn) between layers a and b must be nonzero. The larger the value of Δn, the larger will be the reflectance of the multilayer stack.

This experiment follows a semiempirical, iterative approach to designing and synthesizing the Bragg stack. The two layers that make up the unit cell of the structure must be phase-matched ($n_1 L_1 = n_2 L_2$, where n_i and L_i are the refractive index and the thickness of an individual layer i, respectively) [11]. In order to accomplish this, we will first etch two samples and measure their optical spectrum. These spectra will provide a starting point to calculate the Bragg waveform parameters, because they will allow us to estimate the appropriate time to etch each of the two layers that constitute the unit cell of the Bragg.

First, a single quarter wavelength layer is etched (say $10\,\mathrm{mA\,cm^{-2}}$ for 3 s). The value of nL for this layer is measured, and then a sample is etched using different values for current density and time. The spectrum of this sample is measured and the etch time adjusted until the product of current and time produces a film that has a value of nL matching the first layer. The stopband (λ_S) will be centered at $4nL$ (Equation 3.4), and its width and reflectivity will be maximized if you maximize the index contrast between the two layers (value of n, not nL, roughly proportional to current density).

$$\lambda_S = 4 n_i L_i \qquad (3.4)$$

Our choice of the current density values used to prepare the two layers will be the same as the maximum and minimum current density values used to prepare the rugate filter in Experiment 3.2. We will etch the two test samples for a period of time that is about 10 times longer than the period we will use in our final calculated waveform, to provide more accurate optical thickness values. This experiment requires you to perform optical reflectance measurements on your samples. The measurement is described in Experiment 5.2.

EXPERIMENT 3.3: Bragg stack by programmed electrochemical etch of p^{++}-type silicon wafer

In this experiment you will prepare a 1-D photonic crystal known as a Bragg stack from a highly doped p^{++}-type silicon wafer. This procedure produces a mixed microporous/mesoporous sample that is 15 μm thick, with nominal pore diameters of the order of 10 nm and average porosity of 60%. The spectral resonance appears at approximately 520 nm

Equipment/supplies:

Programmable current power supply	See Table 3.2 for examples. For this experiment, a Kepco ATE 25-2DM power supply was used, operating in voltage-programmed current mode with an analog voltage input from a National Instruments NI PCI-6221 D/A card and a microcomputer running National Instruments' Labview software. The Labview program subroutines are available at http://sailorgroup.ucsd.edu/software
Pt spiral counter-electrode	See Table 2.2. A loop or mesh can also be used
Al foil back-contact	This should be fabricated from heavy gauge aluminum foil – aluminum weighing boats from Alfa Aesar (www.alfa.com) can be cut to size
Standard etch cell	Appendix 1
p^{++} type boron doped (100) Si chip, resistivity 1 mΩ-cm	Siltronix (www.siltronix.com). Measurement of resistivity is performed in Experiment 1.1
3:1 aqueous HF (48%): ethanol solution	See Section 2.5

Procedure:

Single-layer Test Samples

1) Clean the silicon chip for 15 minutes in an ultrasonic bath in isopropanol, then dry it. A more thorough cleaning by etching a sacrificial porous layer (Experiment 2.2) is strongly advised.

2) Mount the chip in the Standard etch cell (Appendix 1) with an aluminum foil back-contact.

3) Add electrolyte consisting of 3:1 (v/v) mixture of 48% aqueous HF and absolute ethanol.

4) Immerse the platinum spiral counter-electrode in the electrolyte. Attach the counter-electrode to the negative (black) lead of the power supply. Attach the positive (red) lead to the aluminum foil back-contact.

5) Set the power supply to current control mode (voltage limit turned all the way up) with the voltage output from the computer DAC card to the "external" voltage input on the power supply. Note the power supply will be operating in a "voltage programmable current" mode; the machine must have a voltage to current conversion circuit built in. You should test that the computer is controlling the current properly using the test configuration shown in Figure 2.2.

6) Activate the power supply and run a constant current density of $12.5\,mA\,cm^{-2}$ for 300 s. This current density corresponds to 15 mA if you are using the Standard etch cell from Appendix 1.

7) After completion of the etch, turn off the power source and remove the HF solution from the cell. Rinse the cell with ethanol, remove the chip and rinse it again with ethanol, then blow it dry under a stream of nitrogen.

8) Measure the reflectance spectrum of the sample and determine the value of nL (see Experiment 5.2).

9) Repeat steps 1–7 with a new silicon chip, using a current density of $90.0\,mA\,cm^{-2}$ (108 mA if you are using the Standard etch cell) for 300 s.

10) Measure the reflectivity spectrum and determine the value of nL for this sample.

Preparation of the Bragg Stack

Before the Bragg stack can be prepared, the duration of the etch for each individual (*a* and *b*) layer in the stack must be determined. Table 3.6 presents the data from the single layer measurements above and the

3.4 More Complicated Photonic Devices

final values of current density and time for each layer *a* and *b* to generate a Bragg centered at 500 nm. The thickness of an individual layer (L) is approximately proportional to the etch duration. Thus, if you need to decrease the value of nL by a factor of five, you simply decrease the duration of the etch by a factor of five. Since the design wavelength for this structure is 500 nm, application of Equation 3.4 yields a value of nL of 125 nm. Rearranging Equation 3.4 gives an expression that can be used to calculate the duration of the etch (t_i) needed for layer i (either layer *a* or *b*) based on parameters we measured on the two test samples we prepared above:

$$t_i = \frac{\lambda_s}{4 n_{test} L_{test}} \times t_{test} \tag{3.5}$$

where λ_s is the desired wavelength of the stop band, n_{test} and L_{test} are the refractive index and the physical thickness of the test layer, respectively, and t_{test} is the etch duration for the test layer. The product $n_{test}L_{test}$ is the optical thickness of the layer, readily determined from an optical reflectance measurement (see Experiment 5.2). Figure 3.9 gives the reflectance spectra for the single-layer test samples.

The test samples yielded values of nL of 4230 and 20 330 nm for layer *a* and *b*, respectively. To satisfy the phase matching condition ($n_1L_1 = n_2L_2$), the etch duration for layer *a* must be decreased by a factor of 4230/125 = 33.8, corresponding to 8.875 seconds (300/33.8). Similarly, layer *b* must be etched for a duration of $(500 \times 300)/(4 \times 20\,330) = 1.845$ s to achieve the appropriate stop band at 500 nm.

The data from Table 3.6 can be used to generate the waveform needed to etch the Bragg structure; the waveform is shown in Figure 3.10. Codes to generate such waveforms, written for the National Instruments Labview program, are available at http://sailorgroup.ucsd.edu/software/.

To prepare the sample:

1) Clean the silicon chip for 15 minutes in an ultrasonic bath in isopropanol, then dry it. A more thorough cleaning by etching a sacrificial porous layer (Experiment 2.2) is strongly advised.

2) Mount the chip in the Standard etch cell (Appendix 1) with an aluminum foil back-contact.

3) Add electrolyte consisting of 3:1 (v/v) mixture of 48% aqueous HF and absolute ethanol.

4) Immerse the platinum spiral counter-electrode in the electrolyte. Connect the anode and cathode to the power supply.

5) Using the programmable power supply, etch the waveform of Figure 3.10 (15 mA for 8.875 s, 108 mA for 1.845 s, repeated 100 times).

Your actual current and time values may differ depending on the outcome of the test sample etches.

6) After completion of the etch, turn off the power source and remove the HF solution from the cell. Rinse the cell with ethanol, remove the chip and rinse it again with ethanol, then blow it dry under a stream of nitrogen.

Results:

The Bragg stack is approximately 18 μm thick, with a 180 nm unit cell. The porosity modulation is between 48 and 68%, with an average porosity of 61%. The reflectance spectrum is shown in Figure 3.11. Note that your results may not exactly match those reported here due to differences in wafer resistivity or laboratory temperature. Chapter 5, Experiment 5.2 provides a detailed description of the procedures to follow and equipment needed to perform the optical measurements. This sample was purposely prepared using the same maximum and minimum current values used in the preparation of the rugate filter sample of Experiment 3.2. Compared with the spectrum of that rugate filter, the stop band peak for the Bragg stack is significantly broader. This is a characteristic of the larger photonic band gap that can be produced with Bragg stacks. In order to prepare an even broader stop band, the maximum current used in the procedure can be increased and the minimum current used can be decreased. Either will increase the difference in porosity (and thus the difference in refractive index) between the two layers.

As with the rugate sample described above, the Fabry–Perot fringes apparent in the spectrum can be reduced if the etch begins and ends with current ramps of ~5 s duration. These ramps serve to produce a gradual change in porosity, and thus refractive index, at the air/porous silicon and the porous silicon/silicon substrate interfaces. These gradual index transitions serve as anti-reflection coatings that reduce the intensity of Fabry–Perot fringes in the spectrum. Starting the etch at a current density close to the electropolishing limit (~500 mA cm^{-2}) and then ramping it down to the starting value of the Bragg, and finishing the etch with a ramp to 0 mA cm^{-2} will produce the desired effect.

3.4.2
Multiple Spectral Peaks-"Spectral Barcodes"

Because the etch current versus time profile directly maps to the refractive index versus depth profile, much more complicated nanostructures and optical spectra can be achieved. If an arbitrary current–time waveform is etched into a wafer, the porous nanostructure that results displays an

Table 3.6 Data used to determine the waveform parameters to etch a Bragg stack.

	Bragg layer a	Bragg layer b
Current density, mA cm^{-2}	12.5	90.0
Etch duration for test sample, s	300	300
Value of nL for test sample, nm	4230	20 330
Layer porosity	48%	68%
Desired value of nL for Bragg layer, nm	125	125
Etch duration needed for layer, s	8.875	1.845

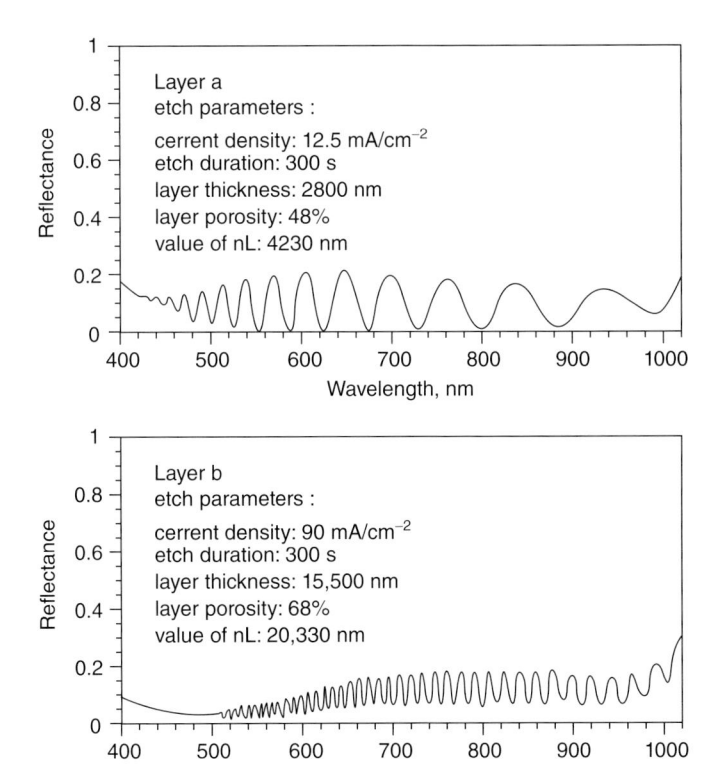

Layer a
etch parameters :
cerrent density: 12.5 mA/cm^{-2}
etch duration: 300 s
layer thickness: 2800 nm
layer porosity: 48%
value of nL: 4230 nm

Layer b
etch parameters :
cerrent density: 90 mA/cm^{-2}
etch duration: 300 s
layer thickness: 15,500 nm
layer porosity: 68%
value of nL: 20,330 nm

Figure 3.9
Reflectance spectra of single-layer test samples, prepared in order to determine the parameters for the waveform needed to etch a Bragg stack. The oscillations in the spectrum are Fabry–Perot fringes that are used to determine the value of nL (product of refractive index and layer thickness). The value of nL for these layers is 4230 and 20 330 nm, respectively. The actual value of nL for each layer (a and b) used in the Bragg structure is 125 nm, (to prepare a sample with a stop band at 500 nm), so the etch duration for each layer must be scaled accordingly (see Table 3.6).

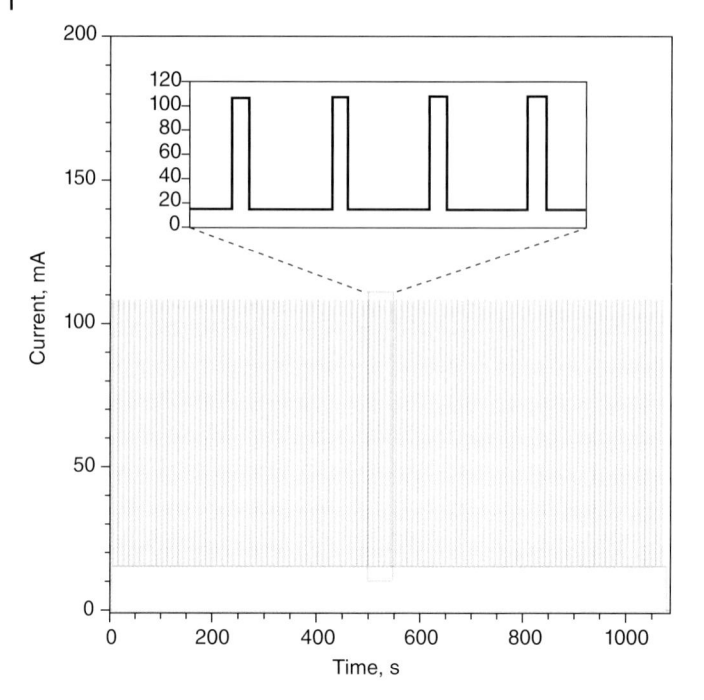

Figure 3.10
Waveform used to prepare a Bragg stack, determined from the analysis summarized in Table 3.6. The waveform consists of a low current density step of 12.5 mA cm^{-2} (15 mA if using the Standard etch cell) for 8.875 s, followed by a high current density step of 90 mA cm^{-2} (108 mA if using the Standard etch cell) for 1.845 s. The waveform has 100 repeats.

optical reflectivity spectrum represented by the Fourier transform of the current–time waveform. The concept was first described by Bovard [13] in the design of optical filters made from vapor deposited dielectrics, and it was first demonstrated in the porous silicon system by Berger et al. [11] The algorithm is outlined in Figure 3.12, and an example of the complicated porosity functions that can be achieved is shown in the cross-sectional electron micrograph image of Figure 3.1. The design and synthesis details are given in the literature [1, 16, 17], and they are summarized briefly here.

A composite waveform f_{comp}, prepared by superimposing n sine waves f_n will yield a porous silicon sample displaying n spectral peaks. The summation can be represented by:

$$f_{comp} = \frac{\sum_n f_n}{n} \tag{3.6}$$

where f_n is a temporal sine wave of the form

$$f_n = A_n[\sin(k_n t - \phi_n) + 1] + A_{n,\min} \tag{3.7}$$

3.4 More Complicated Photonic Devices | 103

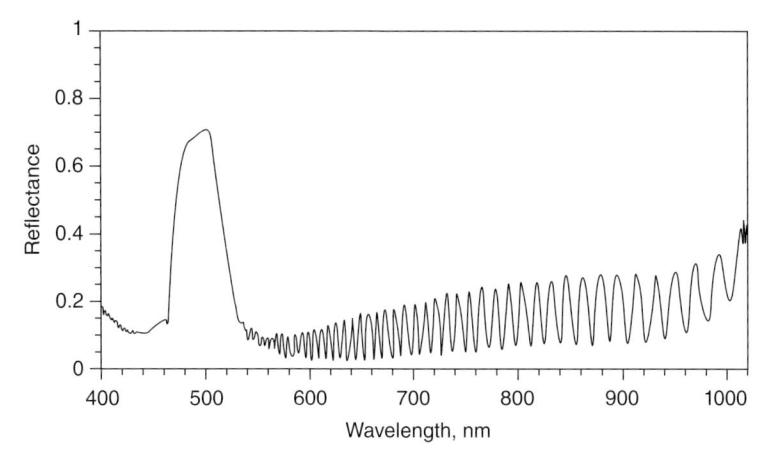

Figure 3.11
Reflectance spectrum of a porous silicon Bragg stack prepared using the etching parameters given in Experiment 3.2. Note that the results reported here are approximate; the spectrum will display a moderate dependence on the resistivity of the wafer used. The oscillations in the spectrum from ~550–1050 nm are Fabry–Perot fringes that correspond to reflection from the front and back sides of the film (at the air/porous silicon and porous silicon/silicon substrate interfaces).

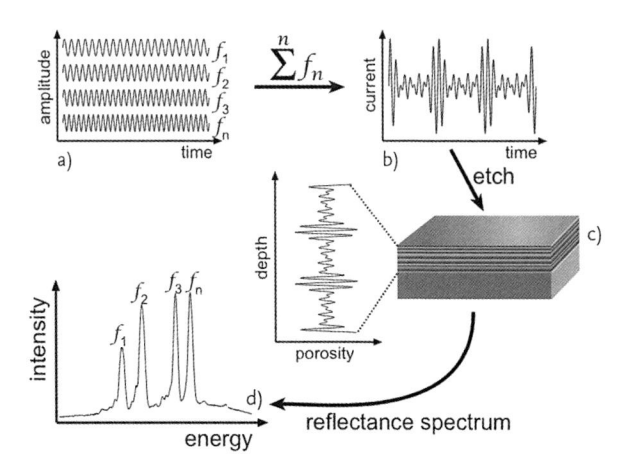

Figure 3.12
Generation of multiple spectral peaks, or "spectral barcodes" by etching porous silicon. Several sine waves with different frequencies (a) are summed to generate a composite waveform (b) that is then converted into a current–time waveform in the computer-controlled current source, etching a porosity–depth profile into the silicon wafer (c). The resulting optical reflectivity spectrum (d) displays the frequency components of the original sine waves as separate spectral peaks. Representing the Fourier transform of the composite waveform, the wavelength and intensity of each peak in the spectrum is determined by the frequency and amplitude, respectively, of its corresponding sine component.

The amplitude (A_n), frequency (k_n), phase offset (ϕ_n), and amplitude offset ($A_{n,\,min}$) are determined by the user, and t is time. The value of A_n for a given component follows Equation 3.8:

$$A_n = (A_{n,\,max} - A_{n,\,min})/2 \qquad (3.8)$$

Where $A_{n,\,max}$ is the maximum amplitude of the sine component n and $A_{n,\,min}$ is the minimum amplitude of the sine component. $A_{n,\,min}$ is the same in both Equations 3.7 and 3.8. All of the individual sine components are summed together to create the composite waveform of Equation 3.6, which is then converted to an analog current–time waveform by the computer-controlled current source, and the single-crystal silicon wafer is then etched. As mentioned above, the optical spectrum (intensity vs. frequency) that results from such a structure represents the Fourier transform of the spatial distribution of refractive index (refractive index vs. distance) in the material (Figure 3.12). Each of the main peaks in the optical reflectance spectrum of this sample corresponds to one of the sine components used in making the composite waveform. Thus the reflectivity spectrum (Figure 3.12d) represents the Fourier transform of the composite current–time waveform (Figure 3.12b) used in preparing the sample.

The ability to "dial-in" a spectrum using the etch waveform superposition method described above represents a simple means to prepare elaborate spectral signatures. The approach has been proposed to be useful for a range of applications, including materials for remote sensing [18], as anti-counterfeiting labels for drugs, explosives, or currency [17], as encoded beads for high-throughput bioassays [16], and as optically probed drug delivery materials [17]. The spectral signatures can be constructed in the visible, NIR, and IR regions of the spectrum [19].

3.5
Lateral Pore Gradients (in the x–y Plane)

Porous silicon films containing a distribution of pore sizes can be electrochemically prepared using an asymmetric electrode configuration such as shown in Figure 3.13. In this arrangement, the potential at the silicon/solution interface varies as a function of distance from the platinum counter-electrode due to the electrical resistance of the solution, leading to a decrease in current density with increasing distance from the counter-electrode [20]. The asymmetric electrode configuration provides a current density gradient across the wafer that results in a distinct variation in the morphology of the pores. For p^+- or p^{++}-type samples, this translates to simultaneous variations in pore diameter, porosity, and film thickness. Representative optical photographs of wafers etched using the asymmetric electrode configuration are displayed in Figures 3.13 and 3.14. The HF(aq):ethanol ratio for the samples

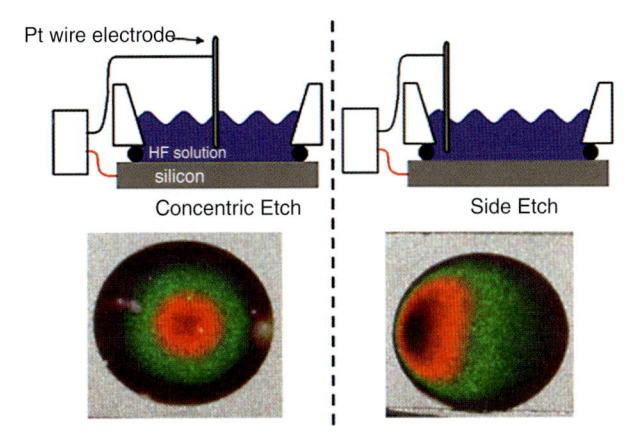

Figure 3.13
Generation of pore gradients by etching porous silicon using an asymmetric counter-electrode configuration. Two examples of rugate filters etched using a thin platinum wire counter-electrode are shown. In the "Concentric Etch" example, the counter-electrode is placed in the center of the etch cell. The electrode is inserted toward an edge of the cell for a "Side Etch". The electrochemical current density at any point on the silicon surface is directly proportional to the distance between that point and the platinum wire counter-electrode. The gradient in color corresponds to a porosity and thickness gradient that is generated by the asymmetric current distribution in the cell. The samples shown were prepared using the Standard etch cell (Appendix 1); both circular samples are 1.2 cm in diameter.

in Figure 3.14 was varied between 1:1 (sample A), 1:2 (sample B) and 1:3 (sample C), which determines the current density at which electropolishing occurs. The region closest to the Pt counter-electrode, corresponding to the highest current density, has a shiny appearance and contains no porous structure. As we discussed in Chapter 1, porous silicon formation only occurs below a critical current density value (J_{EP}) [21]. At current densities larger than the critical value silicon dissolves uniformly, leading to the smooth and shiny appearance of the electropolished region. Close-up atomic force microscopy (AFM) views of the remaining three regions are shown in Figure 3.14b. The region next to the electropolished area is a transitional region consisting of an incompletely electropolished surface film of porous silicon with small (micron-scale) features (Figure 3.14, A1). Next to the transitional region is a small area that appears matted to the eye (Figure 3.14, A2). The electrochemical conditions in this region result in high porosity silicon that is susceptible to cracking and peeling upon drying [22]. Uniform, reflective porous silicon is observed adjacent to this region (Figure 3.14, A3) and extends to the far edge of the film.

For a given electrode geometry the properties of a pore gradient film can be varied by the choice of electrolyte composition [24–26]. For example,

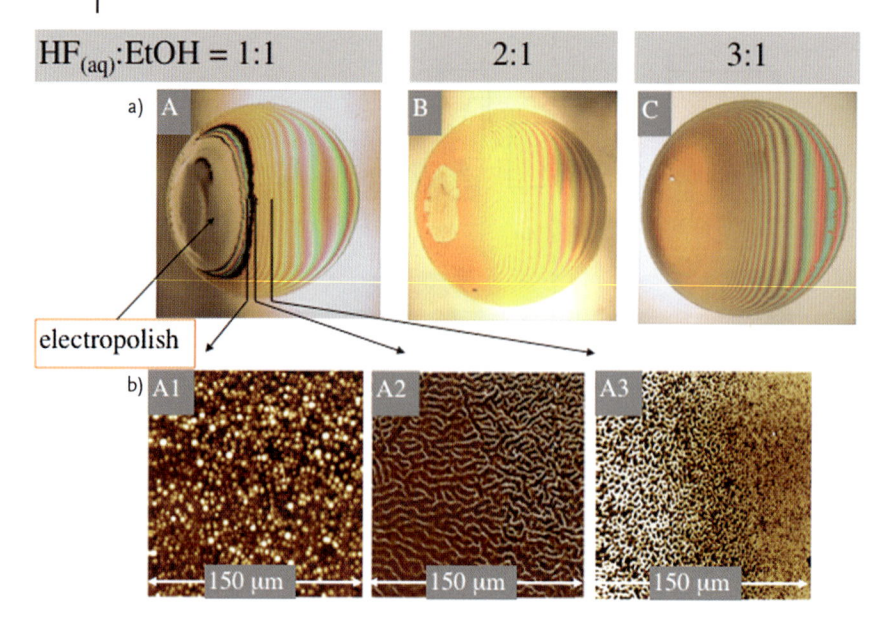

Figure 3.14

Pore gradients generated by etching porous silicon using the "Side Etch" asymmetric counter-electrode configuration of Figure 3.13. (a) All three samples (A, B, and C) were prepared by applying a current density of 115 mA cm^{-2} for 80 s, but different concentration ratios of HF were used in the electrolyte. The specific values of HF(aq):EtOH are as indicated at the top. Each circular sample is 1.2 cm in diameter. (b) The images A1–A3 are tapping-mode atomic force microscope images of various surface morphologies observed on sample A. Pore diameters ranging from ~600 nm (directly adjacent to the electropolished region) to <10 nm (farthest away from the electropolished region) are obtained. Each image window is 150 × 150 mm^2: (A1) very rough region corresponding to the transition between complete electropolish and porous silicon formation (vertical scale 0–400 nm); (A2) matte region (vertical scale 0–1300 nm); (A3) optically smooth region (vertical scale (0–600 nm). The samples shown were prepared using the Standard etch cell (Appendix 1). Images adapted from reference [23].

almost no electropolished region is observed on the silicon sample etched in HF(aq):EtOH = 3:1 solution (Figure 3.14, C), and the range of pore sizes is less pronounced than in samples A and B. In HF(aq):EtOH = 2:1 solutions (Figure 3.14, B), a film with a wider range of pore sizes forms. Pore diameters can range from ~600 nm to <10 nm across the surface of a single sample.

In addition to pore size, both pore depth (film thickness) and porosity also vary with current density across the gradient structure. Because this translates to a variation in optical thickness (product of refractive index and layer thickness), the color of a rugate filter prepared in an asymmetric cell will vary in the lateral dimensions. Two photographs of these so-called "rainbow chips" [27] are presented in Figure 3.13. We generate this type of sample in Experiment 3.4.

EXPERIMENT 3.4: Preparation of a color-graded rugate filter by gradient etch of p^{++}-type silicon

In this experiment you will prepare a 2-D photonic crystal by electrochemical etch of a rugate filter using an asymmetric electrode configuration. The sample will display a narrow spectral resonance whose wavelength varies from one side of the chip to the other, as shown in Figure 3.13.

Equipment/supplies:

Programmable current power supply	See Table 3.2 for examples. For this experiment, a Kepco ATE 25-2DM power supply was used, operating in voltage-programmed current mode with an analog voltage input from a National Instruments NI PCI-6221 D/A card and a microcomputer running National Instruments' Labview software. The Labview program subroutines are available at http://sailorgroup.ucsd.edu/software
Pt wire counter-electrode	Alfa/Aesar (www.alfa.com) platinum wire, 0.5 mm (0.02 in) dia, hard, 99.95% (metals basis) cat number 10286 The wire must be placed in the electrolyte perpendicular to (but not touching) the silicon wafer
Al foil back-contact	This should be fabricated from heavy gauge aluminum foil – aluminum weighing boats from Alfa Aesar (www.alfa.com) can be cut to size
Standard etch cell	Appendix 1
p^{++} type boron doped (100) Si chip, resistivity 1 mΩ-cm	Siltronix (www.siltronix.com). Measurement of resistivity is performed in Experiment 1.1
3 : 1 aqueous HF (48%): ethanol solution	See Section 2.5

Procedure:

1) Clean the silicon chip for 15 minutes in an ultrasonic bath in isopropanol, then dry it. A more thorough cleaning using base removal of a sacrificial electrochemical layer (See Experiment 2.2) is advised.

2) Mount the chip in the Standard etch cell (Appendix 1) with an aluminum foil back-contact.
3) Add electrolyte consisting of 3 : 1 (v/v) mixture of 48% aqueous HF and absolute ethanol.
4) Immerse the platinum wire counter-electrode in the electrolyte. The wire must be placed in the electrolyte perpendicular to (but not touching) the silicon wafer, approximately 3 mm from the wafer face. Attach the counter-electrode to the negative (black) lead of the power supply. Attach the positive (red) lead to the aluminum foil back-contact.
5) Activate the computer-controlled power supply and run the sinusoidal (rugate) waveform described in Experiment 3.2 (cosine wave, 6 s period, 100 repeats, minimum and maximum current densities 12.5 and 90.0 mA cm^{-2}, respectively). If using the Standard etch cell (Appendix 1), the two limiting current density values of 12.5 and 90.0 mA cm^{-2} correspond to 15 and 108mA, respectively.
6) After completion of the etch, turn off the power source and remove the HF solution from the cell. Rinse the cell with ethanol, remove the chip and rinse it again with ethanol, then blow it dry under a stream of nitrogen.

Results:

This procedure follows a published report [27]. The sample should have an appearance similar to the image shown in Figure 3.13. The range of reflectance spectra for a similar sample is given in Figure 3.15. Measurement of the reflectance spectrum is described in Experiment 5.2. Gradient structures such as this have been used for size-dependent molecular separations [23, 28], and as spatially tunable optical filters [27].

3.6
Patterning in the x–y Plane Using Physical or Virtual Masks

The previous experiment gives our first example of generating a pattern in the x–y plane (i.e., across the surface of the polished wafer). This was a fairly low resolution pattern, and the diversity of the features that can be achieved with an electrode-defined current gradient is quite limited. There are a variety of higher resolution approaches that can be used to pattern the surface of a porous silicon film in the x–y plane. The three main approaches are: (i) use of a physical mask, deposited in the form of a photoresist either before or after etching; (ii) a virtual mask based on photo-

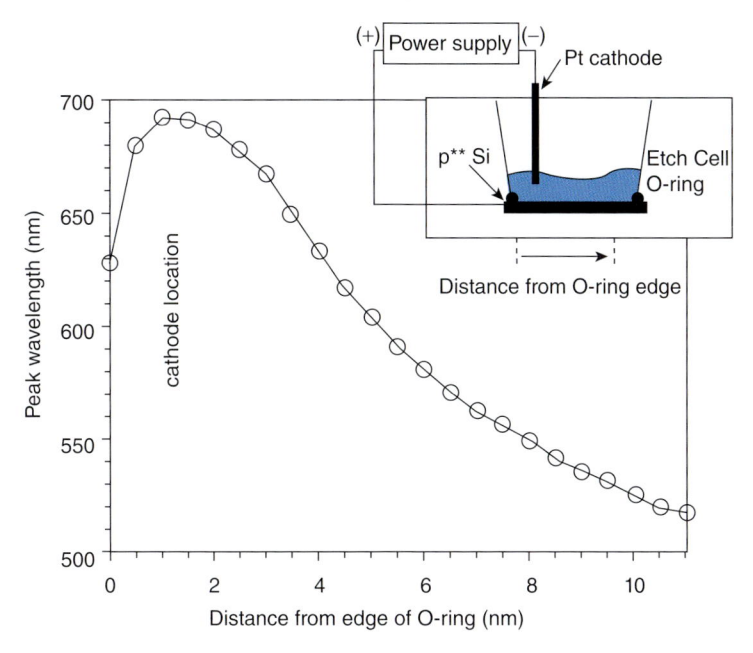

Figure 3.15
Wavelength of rugate peak maximum as a function of distance from the O-ring edge (as defined in the inset) for a porous silicon rugate filter gradient film. Sample was prepared using the "Side Etch" asymmetric counter-electrode configuration of Figure 3.13. Reflectance data were obtained from spectra obtained at normal incidence to the porous silicon film. The sample was etched using a sinusoidal current density waveform oscillating between 11.5 and 34.6 mA cm^{-2}, with 70 repeats and a periodicity of 9 s. The sample was prepared using the Standard etch cell (Appendix 1). Data from reference [27].

electrochemical etching a light pattern projected on the wafer during the etch; and (iii) use of microcontact printing methods, also known as "soft lithography" to mask or remove selected regions of a porous silicon layer. This section briefly outlines the procedures, though their application requires more specialized training or equipment that is not covered in this book.

3.6.1
Physical Masking Using Photoresists

A photoresist can be placed on the wafer either before or after etching of the porous silicon layer, as outlined in Figure 3.16. Either procedure requires the use of conventional photoresist masking and development technologies

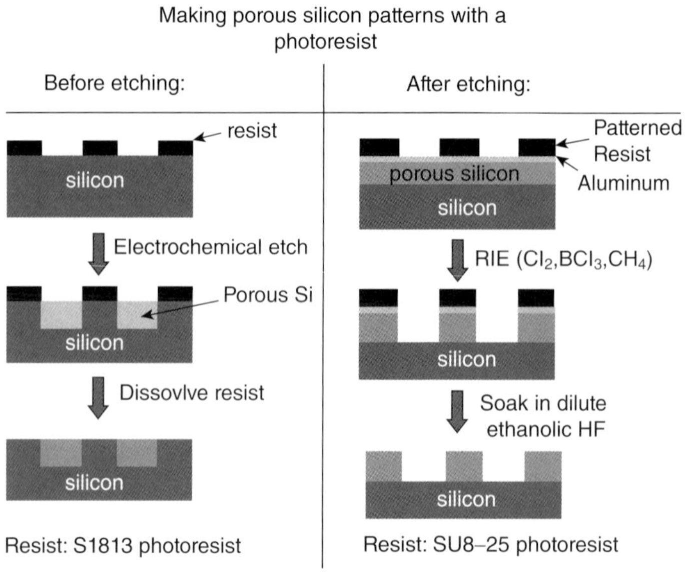

Making porous silicon patterns with a photoresist

Before etching: (Resist: S1813 photoresist)

- Silicon with resist pattern on top
- ↓ Electrochemical etch
- Silicon with Porous Si regions under resist openings
- ↓ Dissolve resist
- Silicon with porous silicon regions

After etching: (Resist: SU8–25 photoresist)

- Silicon coated with porous silicon, Aluminum, and Patterned Resist
- ↓ RIE (Cl_2, BCl_3, CH_4)
- Silicon with patterned porous silicon/aluminum stacks
- ↓ Soak in dilute ethanolic HF
- Silicon with porous silicon regions

Figure 3.16
Two examples of pattern generation using conventional photoresist masks with porous silicon. Details on the "Before etching" approach are given in reference [33]; details on the "After etching" approach can be found in reference [16]. In the "After etching" method, an aluminum layer is deposited on the surface prior to placement of the photoresist, in order to prevent the resist from entering and clogging the pores in the porous silicon layer.

available in most modern cleanroom facilities. The resolution of the method can be in the submicron range, although undercutting of the mask is often a problem when the porous layer is etched after the physical mask is laid down. There are many examples of conventional photolithography methods applied to porous silicon described in the literature [1, 16, 29–32]. An example of a film prepared using the "spectral barcodes" approach described earlier in this chapter, and then formed into disk-like particles using the "after etching" photolithographic method, is shown in Figure 3.17.

Figure 3.18 shows a cross-sectional image of a macroporous sample prepared by masking the sample prior to etching (using the "before etching" photolithographic method shown in Figure 3.16). One of the chief requirements of the "before etching" patterning approach is that the physical mask be impervious to the HF electrolyte that is used in the electrochemical formation of porous silicon. The S1813 photoresist described in Figure 3.16 is adequate, although perhaps the more popular and controlled approach is to use silicon nitride, deposited onto silicon by low pressure chemical vapor deposition through a separate masking step [34]. Patterning of porous

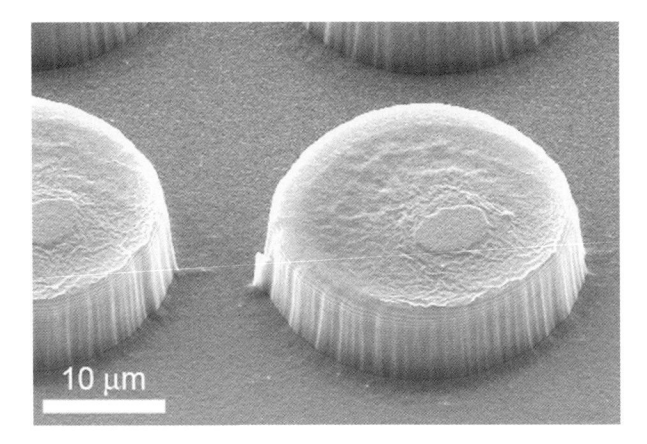

Figure 3.17
Scanning electron microscope image of microfabricated porous silicon photonic crystal particles, prepared using the method outlined in Figure 3.16, by patterning with a photoresist after etching. The particles contain a modulated porous nanostructure that produces a characteristic spectrum, referred to as a "spectral barcode". [16] Image courtesy Shawn O. Meade.

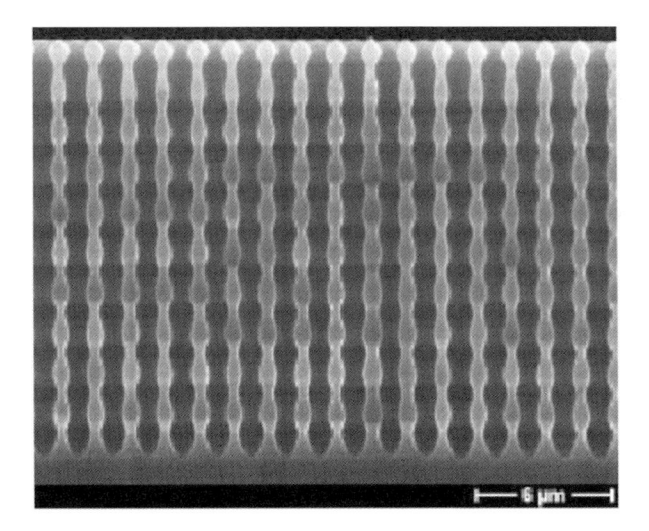

Figure 3.18
Cross-sectional SEM image of n-type macroporous silicon sample prepared by back side illumination during the etch. The spacing of the regular array of pillars (in the x-direction, left to right in the image) is determined by photolithographic masking of the wafer prior to the electrochemical etch. The pore diameter modulation (in the z-direction, top to bottom in the image) is achieved by modulation of the light intensity during the photoelectrochemical etch. From reference [37].

3.6.2
Virtual Masking Using Photoelectrochemistry

Although the sample of Figure 3.18 was etched with the assistance of light (using back side illumination, as described in Experiment 2.4), the patterned, regular array of pillars that is apparent in the image comes from a physical mask that was deposited prior to the photoelectrochemical etch. However, light can be used to control morphology in the x–y plane without the need for a physical mask – in such a case the photovoltaic effect is used to generate valence band holes that locally drive the etch of silicon. Uhlir

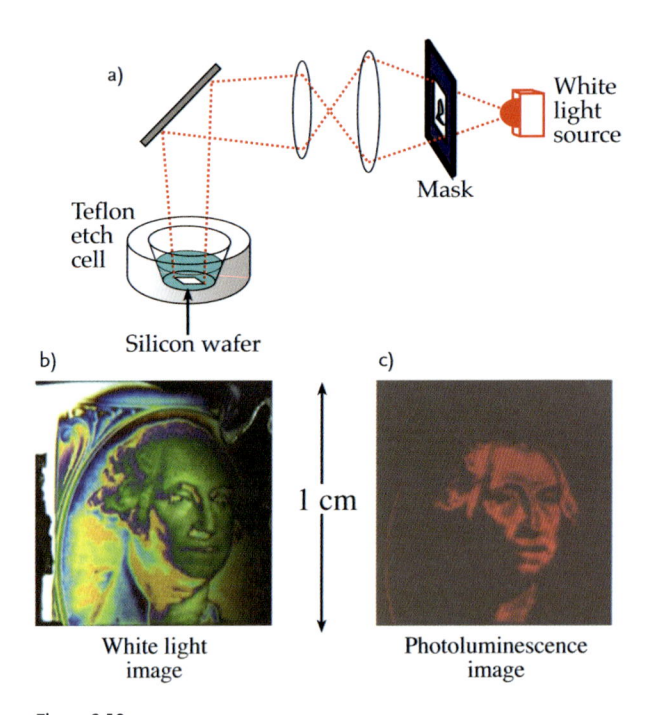

Figure 3.19
(a) Schematic diagram depicting direct photoetching of silicon to make patterns of porous silicon. The etch cell contains a standard HF/ethanol electrolyte, and a relatively low electrochemical current is applied (typically 10 mA cm^{-2}) while an image is projected onto the front face of the silicon wafer. (b) and (c) Two photographs of a sample prepared in this fashion. (b) The appearance of the sample in white light. False colors due to optical interference in the thin layer of porous silicon are seen. (b) The same sample, photographed under ultraviolet light illumination [39, 40].

3.6 Patterning in the x–y Plane Using Physical or Virtual Masks

Table 3.7 Approaches to generate patterned porous silicon.

Method	Description	Reference
Ion beam irradiation	Generates positive patterns in the x–y plane. Ion beams create damage in a crystalline silicon wafer that act as pore nucleation centers for a subsequent electrochemical etch.	[41–45]
Soft lithography	Generates negative patterns in the x–y plane. An elastomer (typically poly dimethylsiloxane, PDMS) is used to deliver a chemical that serves as a physical mask to inhibit porous silicon formation in specified regions. Easier to implement than conventional photolithography.	[46]
Dry removal soft lithography	Generates positive or negative patterns, either freestanding or adhered to a silicon or PDMS substrate. A variant of soft lithography, in which the elastomeric stamp is used to remove porous silicon from specified regions of an already-etched sample. The method relies on the fragility of porous silicon, and its ability to adhere to the silicone stamp.	[47, 48]
Microdroplet patterning	Generates pillars of porous silicon on a silicon substrate. A liquid solution of polymer (polystyrene in toluene) is sprayed through a fine nozzle onto a porous silicon layer. The polymer acts as a physical mask to protect the porous silicon in the regions where a droplet hits the film. The unprotected region of the porous silicon film is dissolved with an aqueous base.	[49]
Scratching the surface	Generates positive or negative patterns in the x–y plane. Physically scribing the surface of a silicon wafer creates defects that nucleate pore formation (positive patterning). If the stylus contains a solution of noble metal ions, local deposition of the metal creates catalytic dissolution sites (positive patterning). If the silicon wafer is immersed in an alkene during scribing, the surface becomes passivated and subsequent electrochemical etch will avoid the regions that have been scratched (negative patterning).	[50–52]
Tandem electrochemical etch, chemical modification	Generates patterns (layers) in the z-direction. This method also takes advantage of the passivating nature of alkenes grafted to the silicon surface. Porous silicon is electrochemically etched, the material is subjected to a chemical hydrosilylation step, and the film is etched a second time. Good for preparing functional particles (see Chapter 6).	[18]

had already recognized this possibility in 1956 and, in his paper describing the first preparation of porous silicon, he provides a method to enhance the rate of photocorrosion by shining light on the silicon wafer during the etch [38]. The process has subsequently been refined to allow reproduction of complex images, diffraction gratings, or patterns of photoluminescent porous silicon [39, 40].

A simplified diagram of the photopatterning experiment is shown in Figure 3.19, along with a wafer that has been patterned with a photoluminescent image from a U.S. dollar. The physics behind the process was described in Chapter 1 (Figures 1.16 and 1.17). To perform this type of etch, an image must be projected onto the wafer during the electrochemical etch. It can be performed with n-type or p-type wafers, generating positive or negative images, respectively. The key requirements for this type of etch are that the wafer possess a relatively high resistivity ($>1\,\Omega\,cm$) to allow sufficient photogeneration of holes, and that the etch cell provide an unobstructed and undistorted optical path for the image. Simple images can be transferred using a digital projector, a reducing lens, and the Standard etch cell fitted with a platinum loop electrode and using a relatively shallow pool of electrolyte, although image quality can be improved if an optical (polystyrene) window contacts the top of the electrolyte and a path is provided for the hydrogen bubbles to escape without accumulating in the optical path.

3.7
Other Patterning Methods

There are many additional methods to pattern porous silicon. Table 3.7 provides an incomplete summary to guide those interested in going further.

References

1 Meade, S.O., and Sailor, M.J. (2007) Microfabrication of freestanding porous silicon particles containing spectral barcodes. *Phys. Status Solidi-Rapid Res. Lett.*, **1** (2), R71–R73.

2 Pacholski, C., and Sailor, M.J. (2007) Sensing with porous silicon double layers: a general approach for background suppression. *Phys. Status Solidi C*, **4** (6), 2088–2092.

3 Pacholski, C., Sartor, M., Sailor, M.J., Cunin, F., and Miskelly, G.M. (2005) Biosensing using porous silicon double-layer interferometers: reflective interferometric Fourier transform spectroscopy. *J. Am. Chem. Soc.*, **127** (33), 11636–11645.

4 Pacholski, C., Yu, C., Miskelly, G.M., Godin, D., and Sailor, M.J. (2006) Reflective interferometric fourier transform spectroscopy: a self-

compensating label-free immunosensor using double-layers of porous SiO2. *J. Am. Chem. Soc.*, **128**, 4250–4252.

5 Orosco, M.M., Pacholski, C., and Sailor, M.J. (2009) Real-time monitoring of enzyme activity in a mesoporous silicon double layer. *Nat Nanotechol.*, **4**, 255–258.

6 Zhang, X.G. (2004) Morphology and formation mechanisms of porous silicon. *J. Electrochem. Soc.*, **151** (1), C69–C80.

7 Ono, S., Saito, M., Ishiguro, M., and Asoh, H. (2004) Controlling factor of self-ordering of anodic porous alumina. *J. Electrochem. Soc.*, **151** (8), B473–BB478.

8 Hall, N. (2003) The photonic opal – the jewel in the crown of optical information processing. *Chem. Commun.*, (21), 2639–2643.

9 Parker, A.R., Mckenzie, D.R., and Large, M.C.J. (1998) Multilayer reflectors in animals using green and gold beetles as contrasting examples. *J. Exp. Biol.*, **201** (9), 1307–1313.

10 Vincent, G. (1994) Optical properties of porous silicon superlattices. *Appl. Phys. Lett.*, **64** (18), 2367–2369.

11 Berger, M.G., Arens-Fischer, R., Thoenissen, M., Krueger, M., Billat, S., Lueth, H., Hilbrich, S., Theiss, W., and Grosse, P. (1997) Dielectric filters made of porous silicon: advanced performance by oxidation and new layer structures. *Thin Solid Films*, **297** (1–2), 237–240.

12 Salem, M.S., Sailor, M.J., Sakka, T., and Ogata, Y.H. (2007) Electrochemical preparation of a rugate filter in silicon and its deviation from the ideal structure. *J. Appl. Phys.*, **101**, 063503.

13 Bovard, B.G. (1993) Rugate filter theory: an overview. *Appl. Opt.*, **32** (28), 5427–5442.

14 Lorenzo, E., Oton, C.J., Capuj, N.E., Ghulinyan, M., Navarro-Urrios, D., Gaburro, Z., and Pavesi, L. (2005) Porous silicon-based rugate filters. *Appl. Opt.*, **44** (26), 5415–5421.

15 Lorenzo, E., Oton, C.J., Capuj, N.E., Ghulinyan, M., Navarro-Urrios, D., Gaburro, Z., and Pavesi, L. (2005) Fabrication and optimization of rugate filters based on porous silicon. *Phys. Status Solidi C*, **2** (9), 3227–3231.

16 Meade, S.O., Chen, M.Y., Sailor, M.J., and Miskelly, G.M. (2009) Multiplexed DNA detection using spectrally encoded porous SiO2 photonic crystal particles. *Anal. Chem.*, **81** (7), 2618–2625.

17 Dorvee, J.R., Sailor, M.J., and Miskelly, G.M. (2008) Digital microfluidics and delivery of molecular payloads with magnetic porous silicon chaperones. *Dalton Trans.*, (6), 721–730.

18 Sailor, M.J., and Link, J.R. (2005) Smart dust: nanostructured devices in a grain of sand. *Chem. Commun.*, 1375–1383.

19 Thonissen, M., and Berger, M.G. (1997) Multilayer structures of porous silicon, in *Properties of Porous Silicon*, vol. 18, (ed. L. Canham), Institute of Engineering and Technology, London, pp. 30–37.

20 Jedamzik, R., Neubrand, A., and Rodel, J. (2000) Production of functionally graded materials from electrochemically modified carbon preforms. *J. Am. Ceram. Soc.*, **83** (4), 983–985.

21 Lehmann, V. (1993) The physics of macropore formation in low doped n-type Si. *J. Electrochem. Soc.*, **140** (10), 2836–2843.

22 Bellet, D., and Canham, L.T. (1998) Controlled drying: the key to better quality porous semiconductors. *Adv. Mater.*, **10** (6), 487–490.

23 Collins, B.E., Dancil, K.-P., Abbi, G., and Sailor, M.J. (2002) Determining protein size using an electrochemically machined pore gradient in silicon. *Adv. Funct. Mater.*, **12** (3), 187–191.

24 Lehmann, V., Stengl, R., and Luigart, A. (2000) On the morphology and the electrochemical formation mechanism of mesoporous silicon. *Mater. Sci. Eng. B*, **B69–70**, 11–22.

25 Nakagawa, T., Sugiyama, H., and Koshida, N. (1998) Fabrication of

periodic Si nanostructure by controlled anodization. *Jpn. J. Appl. Phys.*, **37** (12B), 7186–7189.

26 Zangooie, S., Jansson, R., and Arwin, H. (1998) Microstructural control of porous silicon by electrochemical etching in mixed HCl/HF solutions. *Appl. Surf. Sci.*, **136** (1–2), 123–130.

27 Li, Y.Y., Kim, P., and Sailor, M.J. (2005) Painting a rainbow on silicon – a simple method to generate a porous silicon band filter gradient. *Phys. Status Solidi A Appl. Mater.*, **202** (8), 1616–1618.

28 Karlsson, L.M., Schubert, M., Ashkenov, N., and Arwin, H. (2004) Protein adsorption in porous silicon gradients monitored by spatially resolved spectroscopic ellipsometry. *Thin Solid Films*, **455–456**, 726–730.

29 Kan, P.Y.Y., and Finstad, T.G. (2006) One-step etch-through porous silicon membrane with an open cavity and pore size tuning. *Phys. Status Solidi A Appl. Mater.*, **203** (15), 3743–3747.

30 Loni, A., Canham, L.T., Berger, M.G., Arens-Fischer, R., Munder, H., Luth, H., Arrand, H.F., and Benson, T.M. (1996) Porous silicon multilayer optical waveguides. *Thin Solid Films*, **276** (1–2), 143–146.

31 Lehmann, V., Stengl, R., Reisinger, H., Detemple, R., and Theiss, W. (2001) Optical shortpass filters based on macroporous silicon. *Appl. Phys. Lett.*, **78** (5), 589–591.

32 Meade, S.O., Yoon, M.S., Ahn, K.H., and Sailor, M.J. (2004) Porous silicon photonic crystals as encoded microcarriers. *Adv. Mater.*, **16** (20), 1811–1814.

33 Cunin, F., Schmedake, T.A., Link, J.R., Li, Y.Y., Koh, J., Bhatia, S.N., and Sailor, M.J. (2002) Biomolecular screening with encoded porous silicon photonic crystals. *Nature Mater.*, **1**, 39–41.

34 Nassiopoulou, A.G., Grigoropoulos, S., Canham, L., Halimaoui, A., Berbezier, I., Gogolides, E., and Papadimitriou, D. (1995) Sub-micrometre luminescent porous silicon structures using lithographically patterned substrates. *Thin Solid Films*, **255** (1–2), 329–333.

35 Anderson, R.C., Muller, R.S., and Tobias, C.W. (1994) Porous polycrystalline silicon: a new material for MEMS. *J. Microelectromech. Syst.*, **3** (1), 10–18.

36 Mescheder, U. (2004) Porous silicon: technology and applications for micromachining and MEMS, in *Smart Sensors and MEMS*, (eds S.Y. Yurish and M.T.S.R. Gomes), Springer, Netherlands, pp. 273–288.

37 Matthias, S., Schilling, J., Nielsch, K., Muller, F., Wehrspohn, R.B., and Gosele, U. (2002) Monodisperse diameter-modulated gold microwires. *Adv. Mater.*, **14** (22), 1618–1621.

38 Uhlir, A. (1956) Electrolytic shaping of germanium and silicon. *Bell Syst. Tech. J.*, **35**, 333–347.

39 Doan, V.V., and Sailor, M.J. (1992) Photolithographic fabrication of micron-dimension porous si structures exhibiting visible luminescence. *Appl. Phys. Lett.*, **60** (5), 619–620.

40 Doan, V.V., and Sailor, M.J. (1992) Luminescent color image generation on porous Si. *Science*, **256**, 1791–1792.

41 Schmuki, P., Erickson, L.E., and Lockwood, D.J. (1998) Light emitting micropatterns of porous Si created at surface defects. *Phys. Rev. Lett.*, **80** (18), 4060–4063.

42 Teo, E.J., Mangaiyarkarasi, D., Breese, M.B.H., Bettiol, A.A., and Blackwood, D.J. (2004) Controlled intensity emission from patterned porous silicon using focused proton beam irradiation. *Appl. Phys. Lett.*, **85** (19), 4370–4372.

43 Teo, E.J., Breese, M.B.H., Tavernier, E.P., Bettiol, A.A., Watt, F., Liu, M.H., and Blackwood, D.J. (2004) Three-dimensional microfabrication in bulk silicon using high-energy protons. *Appl. Phys. Lett.*, **84** (16), 3202–3204.

44 Breese, M.B.H., Teo, E.J., Mangaiyarkarasi, D., Champeaux, F., Bettiol, A.A., and Blackwood, D. (2005) Proton beam writing of

microstructures in silicon. *Nucl. Instrum. Methods Phys. Res. B*, **231**, 357–363.

45 Teo, E.J., Breese, M.B.H., Bettiol, A.A., Champeaux, F.J.T., Wijesinghe, T., and Blackwood, D.J. (2007) Tunable colour emission from patterned porous silicon using ion beam writing. *Nucl. Instrum. Methods Phys. Res. B*, **260** (1), 378–383.

46 Harada, Y., Li, X.L., Bohn, P.W., and Nuzzo, R.G. (2001) Catalytic amplification of the soft lithographic patterning of Si. Nonelectrochemical orthogonal fabrication of photoluminescent porous Si pixel arrays. *J. Am. Chem. Soc.*, **123** (36), 8709–8717.

47 Sirbuly, D.J., Lowman, G.M., Scott, B., Stucky, G.D., and Buratto, S.K. (2003) Patterned microstructures of porous silicon by dry-removal soft lithography. *Adv. Mater.*, **15** (2), 149.

48 Gargas, D.J., Muresan, O., Sirbuly, D.J., and Buratto, S.K. (2006) Micropatterned porous-silicon Bragg mirrors by dry-removal soft lithography. *Adv. Mater.*, **18** (23), 3164–3168.

49 Li, Y.Y., Kollengode, V.S., and Sailor, M.J. (2005) Porous silicon/polymer nanocomposite photonic crystals by microdroplet patterning. *Adv. Mater.*, **17** (10), 1249–1251.

50 Peng, K.Q., Hu, J.J., Yan, Y.J., Wu, Y., Fang, H., Xu, Y., Lee, S.T., and Zhu, J. (2006) Fabrication of single-crystalline silicon nanowires by scratching a silicon surface with catalytic metal particles. *Adv. Funct. Mater.*, **16** (3), 387–394.

51 Sieval, A.B., Demirel, A.L., Nissink, J.W.M., Linford, M.R., van der Maas, J.H., de Jeu, W.H., Zuilhof, H., and Sudholter, E.J.R. (1998) Highly stable Si-C linked functionalized monolayers on the silicon (100) surface. *Langmuir*, **14** (7), 1759–1768.

52 Niederhauser, T.L., Lua, Y.Y., Sun, Y., Jiang, G.L., Strossman, G.S., Pianetta, P., and Linford, M.R. (2002) Formation of (functionalized) monolayers and simultaneous surface patterning by scribing silicon in the presence of alkyl halides. *Chem. Mater.*, **14** (1), 27–29.

4
Freestanding Porous Silicon Films and Particles

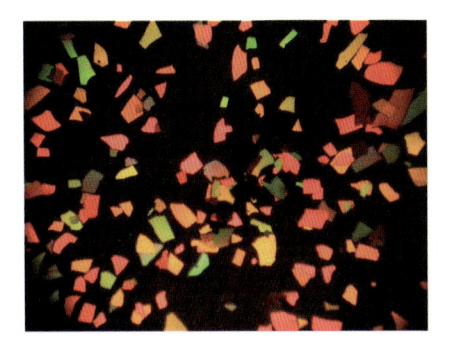

Figure 4.1
A porous silicon layer can be removed from the silicon wafer substrate, generating freestanding films which can be fractured into micron-size or smaller particles. The image shows particles of porous silicon photonic crystals prepared by ultrasonic fracture. Three separate photonic crystal films (rugate filters, see Experiment 3.2) were used to generate the image, each with a different photonic resonance (red, yellow, or green). The particles, roughly 100 µm across, possess the photonic signature of the parent porous silicon film. Image courtesy of Frederique Cunin. From reference ([1]).

This chapter describes the preparation of freestanding films and particles of porous silicon, in which the porous silicon layer has been completely removed from the silicon substrate. The essential requirement for removing a porous silicon layer from the substrate was described in Chapter 1: when the etch current exceeds the rate at which fluoride ions can be transported to the pore tips, the reaction mechanism begins to switch to oxide formation in place of silicon dissolution. This oxide layer eventually spreads to underlay the entire porous layer. Subsequent attack by fluoride ion in the electrolyte dissolves the oxide layer, and a freestanding film of porous silicon results. We refer to this process as "lift-off." Here we will describe the preparation of a lift-off film, and processing steps that can be used to prepare micron-scale or nanometer-scale particles. The experiments in this chapter leave out some of the details that were included in the experiments of Chapters 2 and 3. Refer to those experimental descriptions if you have trouble with any of the procedures described in this chapter.

4.1
Freestanding Films of Porous Silicon-"Lift-offs"

Experiment 4.1 uses a Fabry–Perot layer prepared from a p^{++}-type silicon wafer. However, almost any film prepared in Chapters 2 or 3 can be lifted off with minor modifications of this procedure. The controlling parameters are fluoride ion concentration (which must be relatively low) and current density. The key to the technique is to supply sufficient current such that the rate of delivery of valence band holes to the silicon/porous silicon interface exceeds the mass transfer rate of fluoride ion to that interface. The lack of fluoride ion causes the silicon oxidation reaction to consume water instead, leading to formation of insoluble SiO_2 instead of soluble SiF_4 (see Section 1.6). The oxide layer spreads across the porous silicon/bulk silicon interface, shutting off the electrochemical current as it goes (Figure 4.2). Subsequently, the oxide layer is dissolved by HF, resulting in separation of the porous silicon layer from the substrate.

4.2
Micron-scale Particles of Porous Silicon by Ultrasonication of Lift-off Films

Methods to prepare micron-scale and smaller particles of porous silicon were developed soon after the discovery of photoluminescence from porous silicon [2–6] The particles have been proposed for applications such as drug delivery [5–8], remote sensing [9], and diagnostic biosensing [10]. Additionally, in Chapter 3 we described a method to prepare spectrally encoded particles ("spectral barcodes"), which have been developed for high-throughput bioassays and product surety applications [11–13]. The particles can also be prepared from photoluminescent porous silicon, enabling biomedical imaging applications, both *in vitro* [10] and *in vivo* [14]. They provide attractive alternatives to the more conventional quantum dots derived from II–VI semiconductors, which can be cytotoxic [15] or systemically toxic [16] due to the presence of heavy metal ions in their compositions.

We described methods to use lithography to prepare particles of porous silicon with well-defined shapes in Chapter 3; in Experiments 4.2 and 4.3 we will prepare micro- and nano-scale particles of irregular shape, by ultra-

4.2 Micron-scale Particles of Porous Silicon by Ultrasonication of Lift-off Films | 121

(1) porous Si formation
$[F^-] > [h^+]$

$Si + 6 F^- + 2 H^+ + 2h^+ \rightarrow SiF_6^{2-} + H_2$

(2) oxide formation
$[F^-] < [h^+]$

silicon oxide

$Si + 2 H_2O + 4 h^+ \rightarrow SiO_2 + 4 H^+$

(3) oxide dissolution
$[F^-] > [h^+]$

SiF_6^{2-}
HF

silicon oxide

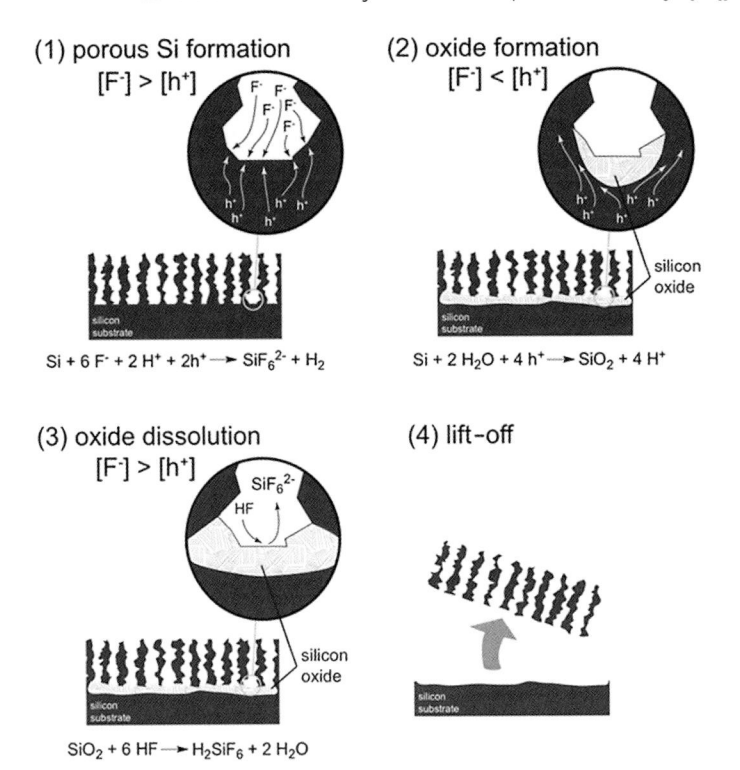

$SiO_2 + 6 HF \rightarrow H_2SiF_6 + 2 H_2O$

(4) lift-off

Figure 4.2
Mechanism for undercutting of a porous silicon layer ("lift-off"). (1) Porous silicon is formed when the rate of fluoride ion transport to the interface is slower than the rate of silicon oxidation. (2) If the supply of fluoride ions is insufficient, the product of the silicon corrosion reaction switches to silicon oxide rather than SiF_6^{2-}, undercutting the porous layer. (3) The oxide is then attacked by HF in the electrolyte, releasing the porous layer as a freestanding film (4).

sonic fracture. The ultrasonication route was the first method used to prepare small particles of porous silicon [2], and it is less expensive and less time-consuming than the lithographic method. The procedure uses porous silicon prepared by electrochemical etching, although porous silicon particles prepared from metallurgical or semiconductor grade silicon powders via stain etching are now commercially available (from Vesta Ceramics, http://vestaceramics.net/). The advantage of the electrochemical route is that it allows the preparation of more complicated morphologies and optical nanostructures, such as multilayers and photonic crystals.

The synthesis of micro- or nanoparticles of porous silicon by the ultrasonication route is outlined in Scheme 4.1. First, the porous silicon layer is etched into a single-crystal silicon substrate in ethanolic HF solution. As with Experiment 4.1, the porous layer is removed from the crystalline

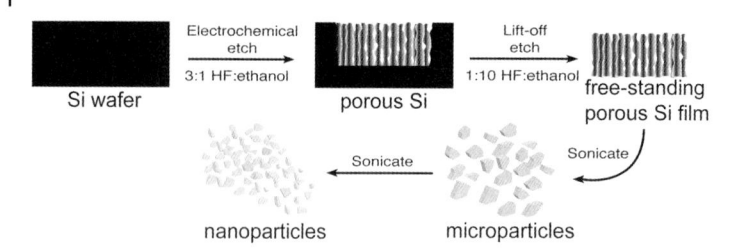

Scheme 4.1

Preparation of micron- and nanometer-size porous silicon particles by ultrasonic fracture [2].

EXPERIMENT 4.1: Freestanding porous silicon film by electrochemical etch of p^{++}-type silicon wafer

In this experiment you will prepare a Fabry–Perot layer of porous silicon from a highly doped p^{++}-type wafer, and then remove it from the substrate to generate a freestanding porous film. This procedure uses the large etch cell. The method produces a mesoporous sample that is ~10 μm thick, with nominal pore diameters of the order of 10 nm.

Equipment/supplies:

Constant current power supply	See Table 3.2 for examples. For this experiment, a Kepco ATE 25-2DM power supply was used, operating in voltage-programmed current mode with an analog voltage input from a National Instruments NI PCI-6221 D/A card and a microcomputer running National Instruments' Labview software. The Labview program subroutines are available at http://sailorgroup.ucsd.edu/software/
Pt spiral counter-electrode	See Table 2.2. A loop or mesh can also be used
Al foil back-contact	Fabricated from heavy gauge aluminum foil – aluminum weighing boats from Alfa Aesar (www.alfa.com) can be cut to size
Large etch cell	Appendix 1. This cell exposes 8.6 cm² of silicon to the etchant
p^{++}-type boron-doped (100) Si chip, resistivity 1 mΩ cm	Siltronix (www.siltronix.com). Measurement of resistivity is performed in Experiment 1.1

3:1 aqueous HF (48%): ethanol solution	See Section 2.5
1:29 aqueous HF (48%): ethanol solution	See above. Adding 1 ml of 48% aqueous HF to 29 ml of ethanol gives a solution that is 2.4% HF by mass.

Procedure:

1) Clean the silicon chip for 15 min in an ultrasonic bath in isopropanol, dry it. A more thorough cleaning by etching a sacrificial porous layer (Experiment 2.2) is advised but not necessary.

2) Mount the chip in the large etch cell (Appendix 1) with an aluminum foil back-contact.

3) Add the electrolyte 3:1 (v/v) of 48% aqueous HF and absolute ethanol.

4) Immerse the platinum counter-electrode in the electrolyte. Attach the counter-electrode to the negative (black) lead of the power supply. Attach the positive (red) lead to the aluminum foil back-contact.

5) Activate the power supply and pass a current density of 90 mA cm^{-2} for 300 s. If using the large etch cell, the corresponding current value is 720 mA.

6) After completion of the etch, turn off the power source and remove the electrolyte solution from the cell. Rinse the cell with ethanol once, and add the 1:29 aqueous HF (48%): ethanol solution.

7) Activate the power supply and pass a current density of 6 mA cm^{-2} for 120 s. If using the large etch cell, the corresponding current value is 48 mA.

8) Rinse the cell carefully with ethanol several times. The film is easily detached from the substrate at this point, so be careful with the rinsings. Remove the chip and rinse it gently with ethanol into a polypropylene or glass beaker. The film should float free. Place the film into a glass beaker containing hexane for a final rinse, remove the excess hexane, and allow the film to dry in air or under a mild vacuum.

Results:

The resulting film will be 10 μm thick, with a porosity of 67% and a total mass of ~5 mg. The purpose of the final rinse with hexane is to minimize the possibility of shattering of the film due to the strong capillary forces and thermal stresses exerted when ethanol evaporates from the pores. The porous silicon film will either blister ("partial lift-off") or completely separate and float to the surface of the etching cell at the beginning of Step 8. Application of minor pressure around the edges of the film (using tweezers or a razor blade) is usually sufficient to remove the film entirely. If the porous layer is not easily removed from the substrate, increase the current density in step 7. Freestanding films tend to be very brittle and they must be handled with care. The strength of the film is improved with lower porosity (lower current density in the initial etch) or thicker films (longer initial etch time). The same wafer can be etched multiple times.

EXPERIMENT 4.2: Photonic crystal porous silicon particles by electrochemical etch of p^{++}-type silicon wafer

In this experiment you will prepare porous silicon particles from an optical rugate filter, starting from a highly doped p^{++}-type wafer. You will remove the photonic crystal from the substrate, generating a freestanding porous film, and then fracture it into particles. This procedure produces mesoporous particles that are ~15 μm thick and ~50 μm on an edge.

Equipment/supplies:

Programmable current power supply	See Table 3.2 for examples. For this experiment, a Kepco ATE 25-2DM power supply was used, operating in voltage-programmed current mode with an analog voltage input from a National Instruments NI PCI-6221 D/A card and a microcomputer running National Instruments' Labview® software. The Labview® program subroutines are available at http://sailorgroup.ucsd.edu/software/
Pt spiral counter-electrode	See Table 2.2. A loop or mesh can also be used
Al foil back-contact	Fabricated from heavy gauge aluminum foil – aluminum weighing boats from Alfa Aesar (www.alfa.com) can be cut to size
Standard etch cell	Appendix 1.

p^{++}-type boron-doped (100) Si chip, resistivity 1 mΩ cm	Siltronix (www.siltronix.com). Measurement of resistivity is performed in Experiment 1.1
3:1 aqueous HF (48%): ethanol solution	See Section 2.5
1:29 aqueous HF (48%): ethanol solution	See above. Adding 1 ml of 48% aqueous HF to 29 ml of ethanol gives a solution that is 2.4% HF by mass.
Ultrasonic bath	Branson Ultrasonic Cleaner, model CPN-952-116 (available from VWR as part #33995-530, www.vwr.com).

Procedure:

Mount the chip in the Standard etch cell (Appendix 1) with an aluminum foil back-contact.

Clean the silicon chip by etching a sacrificial porous layer (Experiment 2.2).

Add the electrolyte 3:1 (v/v) of 48% aqueous HF and absolute ethanol.

Immerse the platinum counter-electrode in the electrolyte. Attach the counter-electrode to the negative (black) lead of the power supply. Attach the positive (red) lead to the aluminum foil back-contact.

Activate the power supply and run the sinusoidal current waveform defined in Figure 3.4: cosine wave, 6 s period, 100 repeats, minimum and maximum current densities 12.5 and 90.0 mA cm^{-2}). If using the Standard etch cell, the two limiting current density values of 12.5 and 90.0 mA cm^{-2} correspond to 15 and 108 mA, respectively.

After completion of the etch, turn off the power source and remove the electrolyte solution from the cell. Do not remove the porous silicon chip from the cell. Rinse the cell with ethanol once, and add the 1:29 aqueous HF (48%): ethanol solution.

Activate the power supply and pass a current density of 6 mA cm^{-2} for 120 s. If using the Standard etch cell, the corresponding current value is 7.2 mA.

Rinse the remaining HF from the cell carefully with ethanol (several rinsings). The film may fall off the substrate at this point, so be careful. If the porous layer is still adhering to the silicon wafer, gently scrape the film free with tweezers or a razor blade and rinse the released film

> into a beaker with ethanol. Place the film in a 20-ml glass vial and add 10 ml of ethanol.
>
> Place a cap on the glass vial. Mount the vial in the ultrasonicator bath with a three-finger clamp. Activate the ultrasonicator for 10 s.
>
> **Results:**
>
> The particles will be ~20 µm thick, with a porosity of 67%. The particles will have an irregular shape, as seen in Figure 4.1. The spectral reflectance peak for the rugate filter appears at ~520 nm. The ultrasonication process will oxidize the particles somewhat; oxidation is more pronounced if the particles are placed in water instead of ethanol, or if the sample is subjected to ultrasonication for a long period of time. The particle size is also very dependent on the amount of time spent in the ultrasonicator [17]. As will be seen in the next experiment, ultrasonication for extended periods of time will result in nanoparticles.

substrate using a lift-off etch in an electrolyte containing a low concentration of HF. The freestanding hydrogen-terminated porous silicon film is then placed in a liquid (pure ethanol is often used, but many liquids, including water, are suitable) and a vial of the mixture is subjected to ultrasound radiation in a common ultrasonic cleaner. The process fractures the porous silicon film into small particles, whose average size can range from a few microns to a few nanometers, depending on the liquid used and the duration of the treatment [17]. The particles can then be centrifuged or filtered if a specific size cut is needed. This experiment demonstrates the preparation of particles from a porous silicon rugate filter, made in a fashion similar to Experiment 3.2.

4.3
Core–Shell (Si/SiO$_2$) Nanoparticles of Luminescent Porous Silicon by Ultrasonication

In Experiment 4.3 you will prepare an aqueous dispersion of luminescent porous silicon nanoparticles starting from a highly doped p^{++}-type wafer. As in Experiment 4.2, you will first generate a freestanding porous film and then fracture it by ultrasonication. The main difference relative to Experiment 4.2 is that you will subject the solution to ultrasonic energy for a significantly longer time (8 h), in order to push the particle size into the nanometer range. We refer to these particles as "core–shell" materials because the luminescent silicon "core" skeleton is encased in a silicon oxide "shell." As with the more common II–VI core–shell nanoparticles [18], the shell passivates the luminescent nanoparticle surface, increasing its radiative efficiency, and thus its photoluminescence intensity. The emissive quantum yield for the particles made by this procedure is ~10%. Unlike the more traditional II–VI "quantum dots," the emission process from this type of material is thought to be a mixture of quantum confinement and defect emission. The defects are associated with the interfacial oxide [10].

Microparticles (Experiment 4.2) and nanoparticles (Experiment 4.3) prepared by ultrasonication of electrochemically etched porous silicon layers can display photoluminescence due to oxide defects and quantum confinement, just like the parent porous silicon material. However, it is important to keep in mind that these structures are typically much larger than the ~2 nm size of an individual quantum dot. Like a Christmas tree, nanoparticles derived from porous silicon consist of a larger silicon superstructure decorated with photoluminescent quantum-confined Si nanocrystalline domains. Figure 4.3 shows porous silicon nanoparticles.

Figure 4.3
Porous silicon nanoparticles.
(a) Photograph of a vial containing porous silicon nanoparticles dispersed in water.
(b) The same vial, illuminated with a UV lamp and showing the distinctive orange glow of luminescent porous silicon.
(c) Electron microscope image obtained from a cluster of dried nanoparticles, each approximately 150 nm in size. The inset shows a close-up view of the pore structure of an individual nanoparticle. Images courtesy of Luo Gu, UCSD [14].

EXPERIMENT 4.3: Luminescent nanoparticles of porous silicon from electrochemically etched p^{++}-type silicon

This procedure produces microporous nanoparticles that are ~150 nm in diameter, with pores of the order of 10 nm.

Equipment/supplies:

Programmable current power supply	See Table 3.2 for examples. For this experiment, a Kepco ATE 25-2DM power supply was used, operating in voltage-programmed current mode with an analog voltage input from a National Instruments NI PCI-6221 D/A card and a microcomputer running National Instruments' Labview software. The Labview program subroutines are available at http://sailorgroup.ucsd.edu/software/
Pt spiral counter-electrode	See Table 2.2. A loop or mesh can also be used
Al foil back-contact	Fabricated from heavy gauge aluminum foil – aluminum weighing boats from Alfa Aesar (www.alfa.com) can be cut to size
Standard etch cell	Appendix 1.
p^{++} type boron doped (100) Si chip, resistivity 1 mΩ-cm	Siltronix (www.siltronix.com). Measurement of resistivity is performed in Experiment 1.1
3 : 1 aqueous HF (48%): ethanol solution	See Section 2.5
1 : 29 aqueous HF (48%): ethanol solution	Adding 1 ml of 48% aqueous HF to 29 ml of ethanol gives a solution that is 2.4% HF by mass.
Ultrasonic bath	Branson Ultrasonic Cleaner, model CPN-952-116 (available from VWR as part #33995-530, www.vwr.com).

Procedure:

1) Mount the chip in the Standard etch cell (Appendix 1) with an aluminum foil back-contact.

4.3 Core–Shell (Si/SiO2) Nanoparticles of Luminescent Porous Silicon by Ultrasonication

2) Clean the silicon chip by etching a sacrificial porous layer (Experiment 2.2).

3) Add the electrolyte 3:1 (v/v) of 48% aqueous HF and absolute ethanol.

4) Immerse the platinum counter-electrode in the electrolyte. Attach the counter-electrode to the negative (black) lead of the power supply. Attach the positive (red) lead to the aluminum foil back-contact.

5) Activate the power supply and pass a current density of 200 mA cm^{-2} for 150 s. If using the Standard etch cell, the corresponding current value is 240 mA.

6) After completion of the etch, turn off the power source and remove the electrolyte solution from the cell. Do not remove the porous silicon chip from the cell. Rinse the cell with ethanol once, and add the 1:29 aqueous HF (48%): ethanol solution.

7) Activate the power supply and pass a current density of 6 mA cm^{-2} for 200 s. If using the Standard etch cell, the corresponding current value is 7.2 mA.

8) Rinse the remaining HF from the cell carefully with ethanol (several rinsings). The film may fall off the substrate at this point, so be careful. If the porous layer is still adhering to the silicon wafer, gently scrape the film free with tweezers or a razor blade and rinse the released film into a beaker using deionized water. Place the film in a 20-ml glass vial and add 10 ml of deionized water.

9) Place a cap on the glass vial. Mount the vial in the ultrasonicator bath with a three-finger clamp. Subject the sample to ultrasonic irradiation for 8 h.

10) Filter the particles through a 0.22-μm filtration membrane (Millipore).

11) Place the colloidal dispersion (in deionized water) in a centrifuge tube and pelletize the nano particles at 14 000 rpm for 30 min. Remove the supernatant (containing silicic acids and smaller nanoparticles) and redisperse the nanoparticles in deionized water.

Results:

The particles will be in the size range 20–200 nm, with a porosity of ~70%. The ultrasonication process will oxidize the particles somewhat; however, their photoluminescence intensity will increase if the materials are left in deionized water at room temperature for 2 weeks or so before you do step 11. Oxidation and dissolution of the particles is very dependent on temperature. If you use a high power ultrasonic bath, be sure to not let the bath temperature rise above room temperature, or the particles may dissolve rather than fracture. The same wafer can be etched multiple times to provide a larger quantity of nanoparticles. Although nanoparticles as small as 2 nm can be prepared by the ultrasonication route, if it is desired to make isolated, solid silicon quantum dots, there are some excellent preparations from molecular precursors that provide higher yields [19, 20].

References

1 Cunin, F., Schmedake, T.A., Link, J.R., Li, Y.Y., Koh, J., Bhatia, S.N., and Sailor, M.J. (2002) Biomolecular screening with encoded porous silicon photonic crystals. *Nat. Mater.*, **1**, 39–41.

2 Heinrich, J.L., Curtis, C.L., Credo, G.M., Kavanagh, K.L., and Sailor, M.J. (1992) Luminescent colloidal Si suspensions from porous Si. *Science*, **255**, 66–68.

3 Wilson, W.L., Szajowski, P.F., and Brus, L.E. (1993) Quantum confinement in size selected, surface-oxidized Si nanocrystals. *Science*, **262**, 1242–1244.

4 Mangolini, L., and Kortshagen, U. (2007) Plasma-assisted synthesis of silicon nanocrystal inks. *Adv. Mater.*, **19**, 2513–2519.

5 Wang, L., Reipa, V., and Blasic, J. (2004) Silicon nanoparticles as a luminescent label to DNA. *Bioconjug. Chem.*, **15**, 409–412.

6 Li, Z.F., and Ruskenstein, E. (2004) Water-soluble poly(acrylic acid) grafted luminescent silicon nanoparticles and their use as fluorescent biological staining labels. *Nano Lett.*, **4**, 1463–1467.

7 Salonen, J., Kaukonen, A.M., Hirvonen, J., and Lehto, V.-P. (2008) Mesoporous silicon in drug delivery applications. *J. Pharm. Sci.*, **97** (2), 632–653.

8 Anglin, E.J., Cheng, L., Freeman, W.R., and Sailor, M.J. (2008) Porous silicon in drug delivery devices and materials. *Adv. Drug Deliv. Rev.*, **60** (11), 1266–1277.

9 Sailor, M.J., and Link, J.R. (2005) Smart dust: nanostructured devices in a grain of sand. *Chem. Commun.*, 1375–1383.

10 Sailor, M.J., and Wu, E.C. (2009) Photoluminescence-based sensing with porous silicon films, microparticles, and nanoparticles. *Adv. Funct. Mater.*, **19** (20), 3195–3208.

11 Meade, S.O., Chen, M.Y., Sailor, M.J., and Miskelly, G.M. (2009) Multiplexed DNA detection using spectrally encoded porous SiO_2 photonic crystal particles. *Anal. Chem.*, **81** (7), 2618–2625.

12 Meade, S.O., and Sailor, M.J. (2007) Microfabrication of freestanding porous silicon particles containing spectral barcodes. *Phys. Status Solidi-Rapid Res. Lett.*, **1** (2), R71–R73.

13 Meade, S.O., Yoon, M.S., Ahn, K.H., and Sailor, M.J. (2004) Porous silicon photonic crystals as encoded microcarriers. *Adv. Mater.*, **16** (20), 1811–1814.

14 Park, J.-H., Gu, L., Maltzahn, G.V., Ruoslahti, E., Bhatia, S.N., and Sailor, M.J. (2009) Biodegradable luminescent porous silicon nanoparticles for in vivo applications. *Nat. Mater.*, **8**, 331–336.

15 Derfus, A.M., Chan, W.C.W., and Bhatia, S.N. (2004) Probing the cytotoxicity of semiconductor

quantum dots. *Nano Lett.*, **4** (1), 11–18.

16 Geys, J., Nemmar, A., Verbeken, E., Smolders, E., Ratoi, M., Hoylaerts, M.F., Nemery, B., and Hoet, P.H.M. (2008) Acute toxicity and prothrombotic effects of quantum dots: impact of surface charge. *Environ. Health Perspect.*, **116**, 1607–1613.

17 Bley, R.A., Kauzlarich, S.M., Davis, J.E., and Lee, H.W.H. (1996) Characterization of silicon nanoparticles prepared from porous silicon. *Chem. Mater.*, **8** (8), 1881–1888.

18 Dabbousi, B.O., Rodriguez-Viejo, J., Mikulec, F.V., Heine, J.R., Mattoussi, H., Ober, R., Jensen, K.F., and Bawendi, M.G. (1997) (CdSe)ZnS core-shell quantum dots: synthesis and characterization of a size series of highly luminescent nanocrystallites. *J. Phys. Chem. B*, **101** (46), 9463–9475.

19 Zhang, X.M., Neiner, D., Wang, S.Z., Louie, A.Y., and Kauzlarich, S.M. (2007) A new solution route to hydrogen-terminated silicon nanoparticles: synthesis, functionalization and water stability. *Nanotechnology*, **18** (9), 095601.

20 Zou, J., Sanelle, P., Pettigrew, K.A., and Kauzlarich, S.M. (2006) Size and spectroscopy of silicon nanoparticles prepared via reduction of $SiCl_4$. *J. Clust. Sci.*, **17** (4), 565–578.

5
Characterization of Porous Silicon

Figure 5.1
Acquisition of optical reflectance spectrum from a porous silicon particle attached to a pharmaceutical pill. The emergence of small, USB-powered spectrometers (an Ocean Optics USB-4000 is shown in the background) and fiber optic probes has greatly simplified the optical characterization of materials.

This chapter describes techniques to characterize the morphology, chemistry, and optical properties of porous silicon films. The three most important morphological characteristics of a porous silicon film are thickness, porosity, and pore size. For optically flat films, thickness and porosity can be characterized by nondestructive optical tests, which are useful as quality control checks to confirm reproducibility of the etching process. The surface chemistry is most easily characterized by infrared spectroscopy, also a nondestructive test. Electron microscopy is an essential tool to quantify pore size and morphology. Adsorption analysis at cryogenic temperatures

Porous Silicon in Practice: Preparation, Characterization and Applications, First Edition.
Michael J. Sailor.
© 2012 Wiley-VCH Verlag GmbH & Co. KGaA. Published 2012 by Wiley-VCH Verlag GmbH & Co. KGaA.

(e.g., the BET, BJH, and BdB methods) yields complementary information on pore size, as well as surface area and a general idea of the shape of the pores. Photoluminescence spectroscopy, both steady-state and time-resolved, are key tools to characterize the quantum-confined forms of the material.

5.1
Gravimetric Determination of Porosity and Thickness

The porosity and thickness of any porous silicon sample can be determined by gravimetric analysis. This is a destructive test that takes advantage of the fact that freshly etched porous silicon dissolves rapidly in basic aqueous solutions. Single-crystal silicon does so also, but much more slowly. Later in this chapter we will describe a complementary, non-destructive method based on spectral measurement of samples immersed in various liquids (spectroscopic liquid infiltration method, or SLIM). The gravimetric method is based upon our definition of porosity P as the ratio of the volume of the pores to the total apparent volume of the film (see Chapter 1):

$$P = \frac{V_{pores}}{V_{total}} \tag{5.1}$$

The gravimetric measurement is performed by weighing the sample before etch (m_1), after etch (m_2), and finally after chemical dissolution of the porous layer (m_3), Figure 5.2. The volume of the pores is assumed to be equal to the volume of silicon that is removed during the electrochemical etch, which is related to the mass of the wafer before (m_1) and after (m_2) etching:

$$V_{pores} = \frac{m_1 - m_2}{\delta_{Si}} \tag{5.2}$$

Where δ_{Si} is the density of elemental silicon. The total apparent volume of the porous film (including voids) can be determined from the mass of the wafer before etching (m_1) and the mass of the wafer after the porous layer has been removed (m_3):

Figure 5.2
Schematic outlining the procedure used to measure thickness and porosity of a porous silicon layer using the gravimetric method. The terms m_1, m_2, and m_3 represent the mass of each sample as indicated.

$$V_{total} = \frac{m_1 - m_3}{\delta_{Si}} \qquad (5.3)$$

Applying Equations 5.2 and 5.3 to 5.1 yields Equation 5.4: [1]

$$P = \frac{m_1 - m_2}{m_1 - m_3} \qquad (5.4)$$

Note that in this calculation density (δ_{Si}) falls out of the expression, and so the method is not dependent on the density of the material being etched. The accuracy of this method relies on careful handling of the sample and the precision of the balance used. The measurement also yields the thickness of the porous layer (W), which *is* dependent on the density of the material and the planar area of the wafer that was exposed to the etching solution:

$$W = \frac{m_1 - m_3}{A \delta_{Si}} \qquad (5.5)$$

where A is the wafer area exposed to HF during the electrochemical etch. The value of δ_{Si} can be taken as $2.33 \, g \, ml^{-1}$.

In Experiment 5.1 we will use the sample prepared in Experiment 2.2 to calculate the thickness and the porosity of the porous layer, though almost any freshly etched porous silicon sample can be used. Aqueous base does not dissolve larger porous silicon structures (e.g., macroporous silicon) very efficiently; in such cases an ultrasonic bath can sometimes help in the removal of the porous layer.

EXPERIMENT 5.1: Gravimetric determination of the porosity and thickness of a porous silicon sample

---

In this experiment you will measure the thickness and porosity of the sample prepared in Experiment 2.2, using the gravimetric method.

Equipment/supplies:

Porous silicon sample (freshly etched)	This method will work with most porous silicon samples as long as they are freshly etched; many chemically modified samples will not dissolve in the KOH solution.
Analytical balance (0.01 mg accuracy)	Sartorius CP225-D (www.sartorius.com)

Aqueous KOH solution	Prepare ~10 ml of 1 M aqueous KOH solution (add 0.56 g of KOH to a 10-ml graduated cylinder and fill to the mark). You can add 1–2 ml of ethanol to this solution to cut the surface tension and improve the dissolution efficiency. Solid KOH (99%) can be obtained from Sigma Aldrich chemicals (www.sigmaaldrich.com)

Procedure:

1) Obtain a freshly etched porous silicon chip (Experiment 2.2 provides an example for a sample derived from p^{++}-silicon). You should have weighed the sample before (m_1) and after (m_2) etching. Handle the sample with Teflon tweezers.

2) Place the sample in a beaker containing 10 ml of the 1 M aqueous ethanolic KOH solution for 10 min. The sample should be agitated (or ultrasonicated) during the procedure. Initially, you should notice significant gas evolution from the porous silicon layer.

3) Rinse the sample with ethanol and dry it thoroughly before obtaining the final mass (m_3).

Results:

The gravimetric data for the sample prepared in Experiment 2.2 are given in Table 5.1:

Applying Equation 5.4, the porosity is calculated as:

$$\frac{0.31433\,g - 0.31169\,g}{0.31433\,g - 0.31090\,g} = 0.77$$

Thus this sample has a porosity of 77%. The thickness of the porous layer is determined from Equation 5.5:

$$\frac{(0.31433 - 0.31090)\,g\,Si}{1.2\,cm^2} \times \frac{1\,cm^3}{2.33\,g\,Si}$$

$= 1.23 \times 10^{-3}\,cm$, or 12.3 μm.

Here we used the exposed area of the porous silicon layer (1.2 cm^2) and the density of crystalline silicon (2.33 g cm^{-3}) to calculate the thickness. The error in the thickness value depends on the reliability of the number used for the exposed area (see below).

With minor modifications to the formulas, this gravimetric method can also be applied to oxidized porous silicon films (porous SiO_2). Reference ([2]) describes the procedure for porous SiO_2 films.

Table 5.1 Gravimetric data used to determine the porosity and thickness of a porous silicon layer.

m_1	0.31433 g
Mass Si chip before etch	
m_2	0.31169 g
Mass Si chip after etch	
m_3	0.31090 g
Mass Si chip after KOH	

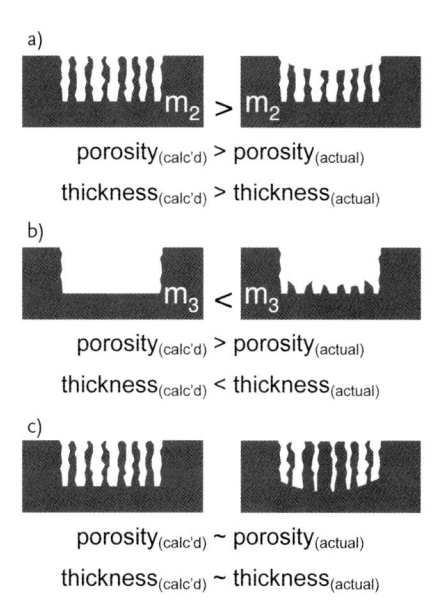

a)

$m_2 > m_2$

porosity$_{(calc'd)}$ > porosity$_{(actual)}$
thickness$_{(calc'd)}$ > thickness$_{(actual)}$

b)

$m_3 < m_3$

porosity$_{(calc'd)}$ > porosity$_{(actual)}$
thickness$_{(calc'd)}$ < thickness$_{(actual)}$

c)

porosity$_{(calc'd)}$ ~ porosity$_{(actual)}$
thickness$_{(calc'd)}$ ~ thickness$_{(actual)}$

Figure 5.3
Three common sources of error in the gravimetric method to determine porosity. The first (a) involves errors in determination of m_2. Thinning of the film during electrochemical preparation is common when etching p-type silicon. The second problem (b) involves error in determination of m_3. Incomplete removal of the porous layer is common with n-type and p^{++}-type porous silicon, due to the relatively large silicon features. The third case (c) depicts a non-uniform porosity or film thickness.

5.1.1
Errors and Limitations of the Gravimetric Method

Gravimetry assumes that the porous silicon film is macroscopically uniform and that the film thickness is the same as the depth of the "crater" left after KOH dissolution. This is not always the case. Three common sources of error are given below (summarized in Figure 5.3):

1) Dissolution of porous silicon causes thinning of the porous silicon layer (Figure 5.3a). The top surface of the porous silicon sample (such as one prepared from p-type silicon) consists of very fine filaments. These can dissolve in the electrolyte during the etching process, reducing the thickness of the layer somewhat. The thinner film yields a smaller than expected value for m_2 in Equation 5.4, and the resulting calculated film porosity will be larger than the actual value.

2) The large (micron-scale) features generated when n-type silicon is etched do not dissolve well in the KOH solution (Figure 5.3b). In this case the value of m_3 will be smaller than anticipated, and the calculated film porosity will be larger than it actually is.

3) Non-uniform porosity or film thickness (Figure 5.3c). A variety of conditions can produce non-uniform porous silicon layers: asymmetric electrode placement generates porosity and thickness gradients in the x–y plane (see Experiment 3.4); poor mixing in the vicinity of the O-ring can lead to depletion of fluoride ion, generating higher porosity and lower thickness values at the edge of the layer; and finally, the HF electrolyte becomes depleted at longer etch times, generating a variation in porosity with depth. Gravimetric measurements provide a geometric average porosity that may be either larger or smaller than that at a given spot in the sample.

5.2
Electron Microscopy and Scanned Probe Imaging Methods

Electron microscopy (transmission electron microscopy and the various scanning electron microscopies) are essential tools to quantify pore size, morphology, and sample thickness. Scanned probe techniques, such as atomic force microscopy and scanning tunneling microscopy are also of some utility for determining pore morphologies. The techniques and relevant instrumentation are ubiquitous, and detailed discussions of the methods are beyond the scope of this book. However, some general pointers for obtaining good images will be mentioned.

5.2.1
Cross-Sectional Imaging

When breaking a wafer to obtain a cross-sectional image, one should be aware that ridges, striations, and other cleavage artifacts can be generated. These can take on the appearance of a porous layer, or the interface between a porous layer and the substrate. It is a good idea to make several measurements along the edge, and to test two or three samples to ensure one is not capturing an aberrant image.

5.2.2
Plan-View (Top-Down) Imaging

As pointed out in Chapter 2 (Figure 2.4), a "crust" layer of microporous silicon often forms during etch of an otherwise mesoporous sample. Plan-view images can thus fool the researcher into a false sense of pore morphology – cross-sectional images are important here. Images of micropores are particularly challenging. Once porosified, even a highly doped silicon sample can become a good insulator. To avoid sample charging under the electron beam, electron microscope images should be obtained at the lowest beam current possible, consistent with the resolution requirements. Sputter-coating a sample with gold is commonly employed to increase conductivity and improve image quality. While this is a satisfactory approach for lower magnification images, if you want to resolve features of ~20 nm or less, sputtered metals will introduce artifacts. Be aware that porous silicon is very fragile. Atomic force microscopy must be performed in intermittent contact (AKA, "tapping mode"), to avoid sample damage during imaging.

5.3
Optical Reflectance Measurements

This section discusses measurements that can be made on smooth, reflecting porous films. We take advantage of the fact that the reflectance spectrum of such porous silicon (or porous SiO_2) films will display an optical interference pattern (Fabry–Pérot fringes) arising from constructive and destructive interference of light from the top and bottom of the porous layer [3]. This occurs when the layer is flat and smooth, with no pores or other features larger than ~500 nm. The pore dimensions in the layer must be significantly smaller than the wavelength of light, or else light scattering will dominate the spectrum and the Fabry–Pérot fringes will not be discernible. The thickness of the film can range from a few hundred nm to >50 µm.

5.3.1
Instrumentation to Collect Reflectance Data

Optical reflectance or transmittance spectra of porous silicon films are readily collected with a spectrometer. Many systems are commercially

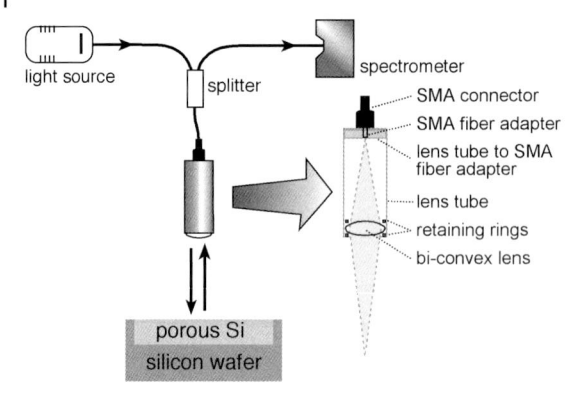

Figure 5.4
Schematic diagram of the optical set-up used to collect reflectance spectra. A CCD spectrometer is fitted with a microscope objective lens coupled to a bifurcated fiber optic cable. A tungsten light source is focused onto the sample surface with a spot size approximately 1–2 mm². Both illumination of the surface and detection of the reflected light are performed along an axis coincident with the surface normal. The parts and vendors for this set-up are listed in Table 5.2.

available; the instrument of choice in this book is an Ocean Optics CCD-based spectrometer, with fiber optic attachments to allow signal input. The heart of the CCD detector is an array of silicon photodetectors fabricated on a monolithic silicon wafer, similar to the light sensors used in digital cameras. Light entering the device is dispersed across the photodetectors by means of a diffraction grating, and the captured image is converted to a digital wavelength versus intensity plot. A system is outlined in Figure 5.4, with the relevant parts listed in Table 5.2.

5.3.1.1 Reflectance Optics

The Ocean Optics USB-4000 CCD spectrometer comes configured with an SMA-type fiber optic input port to receive the light signal. An objective lens coupled to a bifurcated or split fiber optic cable is used to both illuminate and collect the light reflected from the sample. The light source consists of a tungsten filament illuminator. This incandescent source is preferred over a white LED because it has more intensity in the red and near-infrared region of the spectrum, providing a relatively broad effective wavelength range of 400–1000 nm. Both illumination of the surface and detection of the reflected light are performed along an axis coincident with the surface normal. The split optical fiber and lens set-up shown in Figure 5.4 allows illumination and collection of reflected light from a spot on the sample approximately 1 mm in diameter. It is important that the sample be very nearly perpendicular to the spectral axis of the spectrometer/lens assembly.

Table 5.2 Apparatus for collection of optical reflectance data.

Item	Vendor	Part #	Comments
Miniature fiber optic spectrometer, preconfigured for VIS-NIR (350–1000 nm)	Ocean optics (www.oceanoptics.com)	USB4000	USB-powered, 3176 horizontal pixel silicon CCD-based spectrometer.
Split optical fiber: 400 µm splitter fiber, VIS/NIR, 2 m	Ocean optics (www.oceanoptics.com)	SPLIT400-VIS/NIR	The fiber allows illumination and collection along nearly the same optical axis. The fiber core diameter determines the spot resolution and the amount of light collected; the 400 µm fiber diameter is a reasonable compromise for most optical reflectance experiments.
Tungsten halogen light source VIS-NIR (360 nm–2 µm)	Ocean optics (www.oceanoptics.com)	R-LS-1	
Microcomputer	www.dell.com		Almost any computer that has a USB port will work. A notebook computer provides portability and ease of set-up.
SM1 (lens tube) to SMA (fiber) adapter	Thorlabs (www.thorlabs.com)	SM1SMA	
Lens tube, 1 in diameter, 2 in long	Thorlabs (www.thorlabs.com)	SM120	
Bi-convex lens (25.4 mm diameter, 50 mm focal length, BK7 material)	Thorlabs (www.thorlabs.com)	LB1471	
Retaining ring for AE1″ Lens Tube (2 needed), 0.08 Thick	Thorlabs (www.thorlabs.com)	SM1RR	
SMA fiber adapter	Thorlabs (www.thorlabs.com)		

This ensures that the maximum amount of specularly reflected light enters the collection lens, and it avoids the need to correct for the dependence of the reflected spectrum on source–sample–detector angle.

5.3.1.2 Wavelength Calibration

Many commercial CCD spectrometers come with factory-set wavelength calibration. This is true of the Ocean Optics USB-2000 and 4000 CCD spectrometers. However, file corruption or an accidental drop can dislodge this calibration, and so it is a good practice to occasionally check the wavelength calibration of the system. A simple neon lamp, such as those used in instrument displays, can serve the purpose. The spectrum of a small AC lamp, with a few of the more prominent emission lines, is given in Figure 5.5.

The intensity of the signal that is measured by the CCD-based spectrometer is proportional to incident energy. This is because the intensity of the signal at a given wavelength that is output by the device derives from indi-

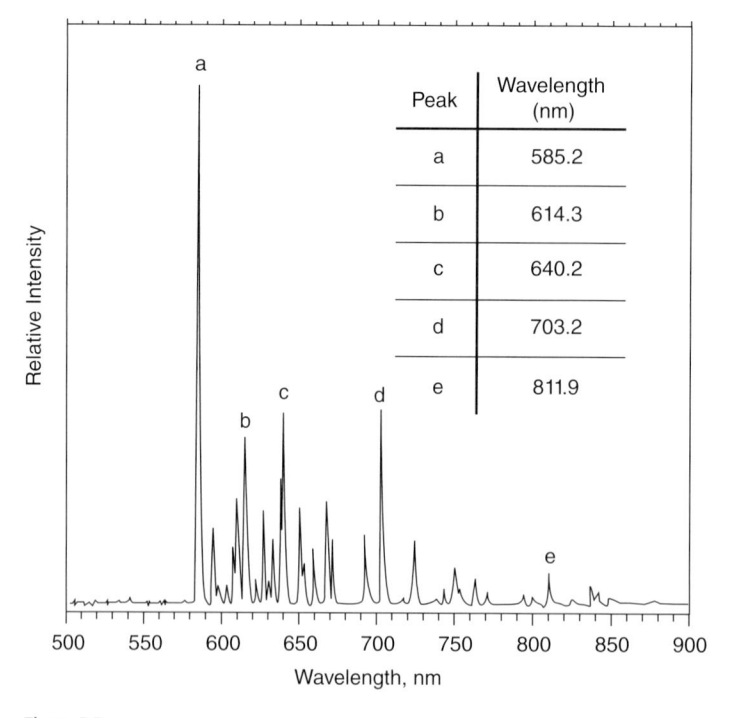

Peak	Wavelength (nm)
a	585.2
b	614.3
c	640.2
d	703.2
e	811.9

Figure 5.5
A spectrum from a Radio Shack inc. 120 VAC neon lamp (#272-707), obtained on an Ocean Optics USB-4000 spectrometer. Wavelength values shown were obtained from the Wavelength table for Ne gas (CRC, 58th Ed., E-210). The strongest emission line is at 585.2 nm.

vidual photons being absorbed in one of the CCD bins positioned at that wavelength. Photon absorption generates an electron–hole pair, and the electrons are counted when the CCD chip is read out. Thus the signal amplitude is directly proportional to the number of photons striking the bin during the exposure period. The number of photons is proportional to the incident energy, and so the amplitude of the CCD signal is directly proportional to incident energy. The exposure time is a user-selectable parameter, so if one wants to increase the incident energy, one just increases the exposure time for the spectrometer chip. To make absolute comparisons between two spectra (for example, measurement of an absolute reflectance spectrum in Experiment 5.2 requires the ratio of a sample and a mirror spectrum), the spectra must be acquired using the same integration time. If the spectral integration time is held constant and the optical geometry is not changed, then the CCD spectra for two or more samples will be proportional to the photon flux (photon s^{-1}), or incident power.

5.3.2
Principles of Fabry–Pérot Interference

Here we will provide a simple derivation of the optical phenomenon known as Fabry–Pérot, or thin-film interference. More detailed derivations can be found in the literature [3, 4]. The derivation here will provide the basic equations needed to understand the SLIM and RIFTS (reflectometric interference Fourier transform spectroscopy) methods used to characterize porous optical films later in this chapter.

When a beam of light strikes a surface, it is either reflected, scattered, transmitted (refracted), or absorbed. For the time being we will assume scattering and absorption are insignificant. We also assume the angle of incidence is normal to the surface, so we can ignore the angular dependence of reflection (Euclid's law of reflection) and of refraction (Snell's law). The situation is depicted in Figure 5.6.

Reflectance "R" is defined as the ratio of the reflected energy (or power) to incident energy at the surface of a structure [5]. For a Fabry–Pérot film, the fraction of light reflected from the surface depends on the index contrast at the interface; reflectance (R) can thus be defined as:

$$R = \frac{\text{reflected power}}{\text{incident power}} = \left(\frac{n_t - n_i}{n_t + n_i}\right)^2 \tag{5.6}$$

where n_i and n_t are the indices of refraction for the two media on either side of the reflecting interface, as defined in Figure 5.6.

The fraction of light transmitted through this interface (transmittance) is given by:

$$T = \frac{\text{transmitted power}}{\text{incident power}} = \frac{4n_t n_i}{(n_t + n_i)^2} \tag{5.7}$$

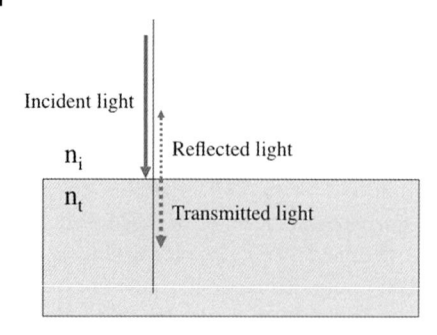

Figure 5.6
Reflection of light from an interface at normal incidence ($\theta_{incident} = \theta_{reflected} = 0$).

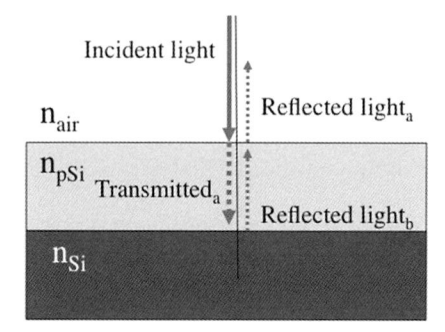

Figure 5.7
Reflection of light from the two interfaces in a porous silicon film.

Note that the quantity $(R + T) = 1$ for all values of n_i and n_t.

Normal incident light is reflected and transmitted at each interface, as long as $n_i \neq n_t$. Using, as an example, a transparent porous silicon layer on a silicon substrate measured in air (Figure 5.7), there are two reflective interfaces to consider: the air/porous silicon interface, and the porous silicon/bulk silicon interface. We can define the index contrast at the two interfaces as ρ_a and ρ_b, given by:

$$\rho_a = \frac{n_{air} - n_{layer}}{n_{air} + n_{layer}}, \quad \rho_b = \frac{n_{layer} - n_{Si}}{n_{layer} + n_{Si}} \tag{5.8}$$

where n_{air}, n_{layer}, and n_{Si} are the refractive indices of air, the porous silicon layer, and bulk silicon, respectively. It follows from Equation 5.6 that the reflectance at each interface is $R(\text{air/porous Si interface}) = \rho_a^2$ and $R(\text{porous Si/Si interface}) = \rho_b^2$. Note that the value we use for the refractive index of the porous silicon layer (n_{layer}) is an "effective" index that represents both the silicon skeleton and the void space – it must take into account the porosity of the composite film. This term is thus an average comprised of

EXPERIMENT 5.2: Measurement of the absolute reflectance spectrum of a porous silicon Fabry–Pérot layer

In this experiment you will measure the absolute reflectance spectrum of a porous silicon sample. A simple Fabry–Pérot layer is used in this example.

Equipment/supplies:

Porous silicon Fabry–Pérot film	Sample for this experiment was prepared following the preparation of "layer b," one of the Fabry–Pérot test layers prepared for the Bragg stack in Experiment 3.3.
Optical spectrometer and optics	See Table 5.2
Broadband metallic mirror	Newport (www.newport.com) model 10D20ER.2, 25.4 mm diameter front-surface silver mirror on a Pyrex® glass support.

Procedure:

1) Etch a porous silicon chip (1 mΩ cm silicon wafer, 3:1 48% aqueous HF:ethanol, 90 mA cm^{-2}, 300 s). Experiment 3.3 provides a more detailed preparation for a sample derived from p^{++}-silicon.

2) Place the sample on a flat surface and then carefully stack the mirror on top, as indicated in Figure 5.8a. Align and focus the optic to a small (~1 mm) spot. Ensure that the optical axis is perpendicular to the mirror.

3) Record the spectrum of the mirror. Be careful not to set the integration time so long that the signal saturates. For this spectrometer, a typical integration time on such a highly reflective surface is 20 ms. Several (~50) scans should be averaged to reduce the noise.

4) Swap the positions of the porous silicon wafer and the mirror, placing the chip on top of the mirror. Be careful not to scratch the mirror surface.

5) Record the spectrum of the sample. The integration time and number of scans should be the same as you used with the mirror spectrum.

Results:

Reflectance Spectrum

The raw spectra (uncorrected for instrumental response) for the mirror and the sample are shown in Figure 5.8a,b, respectively. Dividing spectrum (b) by spectrum (a) yields the spectrum (c). The spectrum generated in this fashion is properly referred to as a reflectance spectrum. Often in the literature one sees the terms reflectance and reflectivity used interchangeably (and incorrectly). Reflectivity "r" is defined as the ratio of amplitudes of the incident and reflected electric field vectors, whereas reflectance "R" is the ratio of the incident to the reflected energy. The two quantities are related by $R = r \times r^*$, where r^* is the complex conjugate of r [5]. Since energy = power × time, if exposure time is held constant then reflectance is also equal to the ratio of the incident to the reflected power, as was indicated by Equation 5.6.

Instrument Response Characteristics

Although the mirror has a fairly flat spectral response, the spectrum of a "white" light source, as recorded by the CCD spectrometer, will not be flat (Figure 5.8a). The shape of the spectrum is a convolution of the response characteristics of the instrumental components: the light source, the silicon detector, the dispersive grating in the spectrometer, and the optics. The decrease in intensity observed in the blue end of the spectrum of the mirror in Figure 5.8a (wavelength range 500 to 400 nm) comes from the lack of blue and UV light output by the tungsten lamp. On the other end of the spectrum, the decrease in intensity observed in the near-infrared region (wavelength range 700 to 1050 nm) comes from the grating response and the lack of sensitivity of the silicon detector as the wavelength of light approaches the bandgap of silicon (1100 nm). The fine structure in the spectrum derives from the grating and the coatings used in some of the optical elements. These features will be present in both the mirror and in the sample spectrum, so they are removed when one spectrum is divided by the other, as seen in Figure 5.8c.

Absorption Losses and the Effective Spectral Window

Comparison of the experimental with the calculated reflectance spectrum (Figure 5.8c,d) reveals deviations associated with the sample and also with the quality of the data. Notice how the amplitude of the oscillations in the reflectance spectrum of Figure 5.8c is attenuated at wavelengths shorter than 650 nm relative to the ideal spectrum. This arises from absorption of light by the porous silicon sample. The absorbance due to the direct bandgap of silicon begins to be noticed at ~650 nm and becomes very strong ($\varepsilon > 0.4$) at wavelengths shorter than 400 nm (Figure 5.9). Absorption losses due to the indirect gap of silicon are relatively minor in the wavelength range 600 to 1000 nm for films less than 50 µm thick. Silicon oxide has very low absorption in the range 400 to 1000 nm, so oxidized porous silicon films tend not to display this feature. At the other end of the spectrum, noise associated with low signal strength causes deviations in the experimental spectrum at wavelengths >1000 nm. Thus, for a typical CCD spectrometer using a tungsten filament light source, the operational spectral window for unoxidized porous silicon samples is 600 to 1000 nm. Although it depends on the degree of oxidation, the useful spectral window for oxidized films can begin as low as 400 nm and extend to 1000 nm.

Analysis of Fabry–Pérot Fringes

The reflectance spectrum of Figure 5.8c displays a series of interference fringes that correspond to constructive and destructive interference of light reflected from the two interfaces present in the structure, referred to as Fabry–Pérot interference [3]. The solutions of Equations 5.9 and 5.10 for the wavelengths of the fringe maxima are given by:

$$m\lambda_{max} = 2nL \qquad (5.11)$$

where m is an integer corresponding to the spectral order of the fringe, L is the thickness of the porous silicon layer, n is the average refractive index of the layer (equivalent to n_{layer} in Equation 5.10), and λ_{max} is the wavelength of the fringe maximum. The factor of 2 derives from the 90° backscatter configuration of the illumination source and detector (Figure 5.4). The term $2nL$ is thus the optical path length, sometimes referred to as the effective optical thickness (EOT).

As expected from the relationship of Equation 5.11, the series of Fabry–Pérot fringes observed in the spectrum of a porous silicon layer are spaced evenly in frequency. Figure 5.10 replots the raw spectrum of Figure 5.8b in terms of the common frequency units, wavenumbers. If each peak in the spectrum is numbered successively, beginning with the lowest energy peak, a plot of that number versus v_{max}, where v_{max} is the frequency of each peak, yields a straight line whose slope is equal to the quantity $2nL$ (Figure 5.10). The correct spectral order of a given fringe (m in Equation 5.11) can be obtained by adding the arbitrarily assigned fringe number to the intercept of the line. For example, the peak labeled "1" in Figure 5.10 is actually the spectral order $m = 41$, "2" is order $m = 42$, and so on.

The plot of Figure 5.10 is only linear if the refractive index is constant; the fact that n for silicon increases sharply for frequencies $> 17000\,cm^{-1}$ (wavelengths <590 nm, Figure 5.9) is manifested as the positive deviation of the data from the linear fit at these higher frequencies. Thus the relationship of Equation 5.11 works well in the wavelength range 600 to 1000 nm if we are working with an unoxidized porous silicon layer. To fit the entire spectral range recorded in Figure 5.8 most accurately, the frequency dispersion of silicon's refractive index and its absorption spectrum (i.e., the complete dielectric function) needs to be taken into account. These codes are readily available; the commercial package SCOUT (www.wtheiss.com) is a very useful tool in this regard.

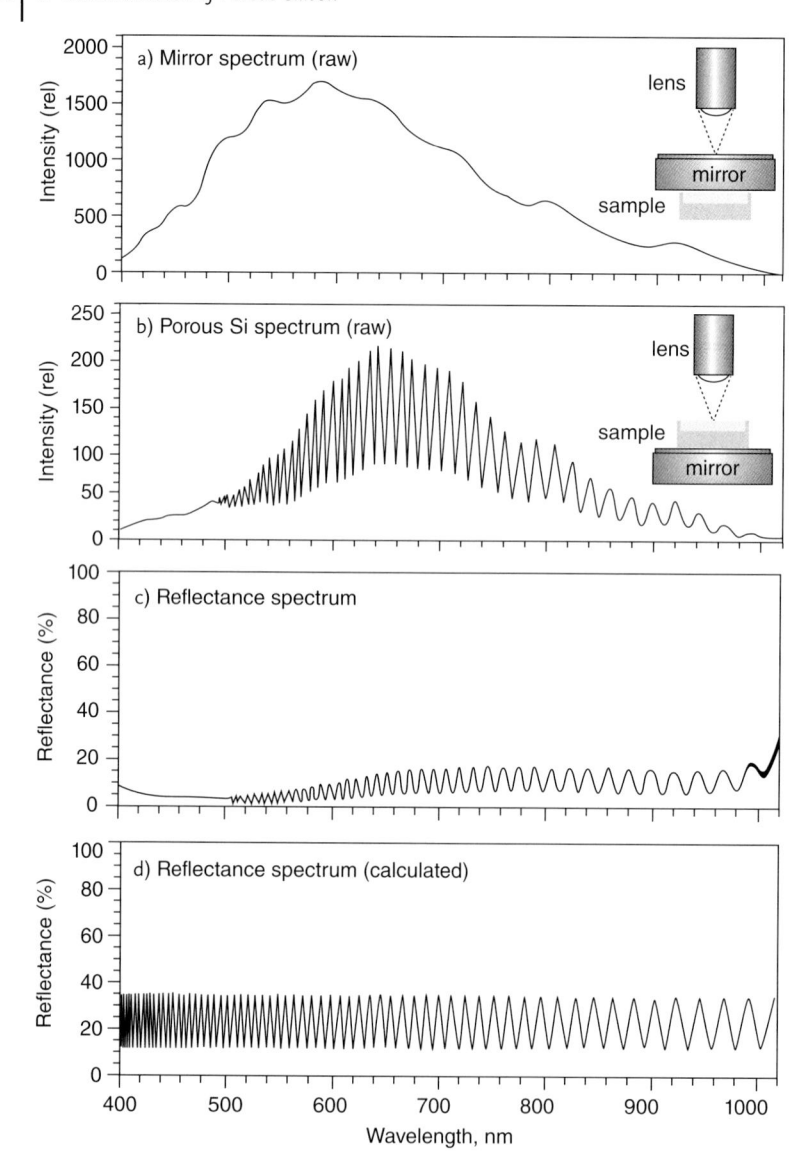

Figure 5.8

"Raw" reflected white light spectra from a mirror (a) and a porous silicon sample (b), used to determine the absolute reflectance spectrum. The sample and mirror are stacked as indicated at the right of each raw spectrum to ensure that both spectra are obtained at the same focal distance. The ratio of the two spectra yields the reflectance spectrum shown in (c). A calculated reflectance spectrum for an ideal Fabry–Pérot film on a silicon substrate (16 µm thick, refractive index 1.27) is shown in (d).

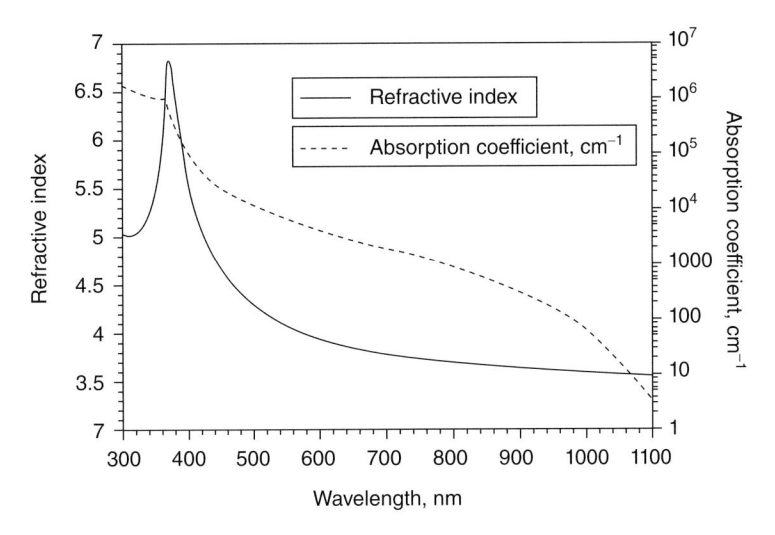

Figure 5.9
Refractive index and absorption coefficient for silicon. Taken from reference ([8]).

the refractive indices of silicon and air. The average is not a simple porosity-weighted average; an effective medium approximation must be used to calculate the index correctly. Many effective medium approximations exist – the Bruggeman [6] or Looyenga [7] models are often applied to porous silicon films. We will discuss effective medium models later in this chapter, in the section on SLIM.

When two reflected beams overlap, the light waves may interfere constructively or destructively, which brings the wave nature of light into the analysis. Thus there will be two sets of terms in the reflectance equation, one for the intensity of reflected light at each interface, and one for the constructive and destructive interference of the two reflected beams:

$$R = (\rho_a + \rho_b) + 2\rho_a\rho_b \cos(2\delta_{layer}) \qquad (5.9)$$

where ρ_a and ρ_b are as defined in Equation 5.8, and the term δ_{layer} represents the phase relationship of the interfering beams in the porous layer, given by:

$$\delta_{layer} = \frac{2\pi n_{layer} L}{\lambda} \qquad (5.10)$$

where n_{layer} is the average refractive index of the porous silicon layer, L is the physical thickness of the porous silicon layer, and λ is the wavelength of light. Note that we are ignoring absorption of light by the porous silicon layer. A more thorough analysis should take absorbance, as well as the wavelength dependence of the refractive index into account.

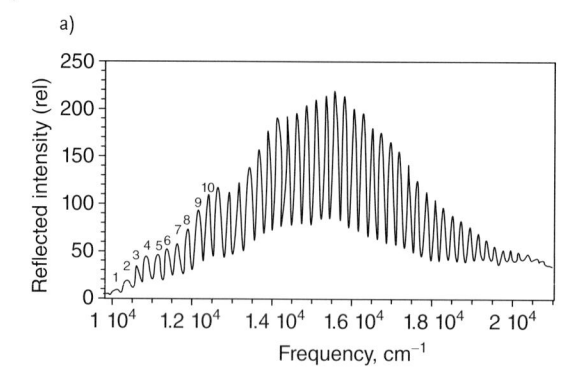

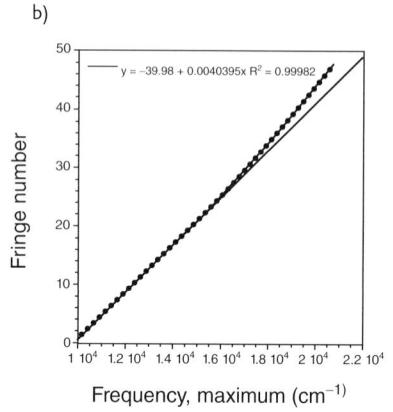

Figure 5.10
(a) "Raw" reflected white light spectrum of the porous silicon sample from Figure 5.8b, plotted in terms of wavenumbers. The first ten fringes observed in the spectrum are numbered. (b) Plot of the integers corresponding to all the fringes as a function of the frequency at which each fringe appears in the spectrum.

Experiment 5.2 walks you through the acquisition and analysis of a typical Fabry–Pérot spectrum. In order to acquire a corrected reflectance spectrum, we need to obtain two spectra: one of the sample and the other of a highly reflective mirror. We then divide the sample spectrum by the mirror spectrum to obtain the absolute reflectance spectrum.

5.3.3
Analyzing Fabry–Pérot Interference Spectra by Fourier Transform: the RIFTS Method

Double- and multi-layer films display more complex fringe patterns that cannot be easily modeled, particularly if one does not know the structure

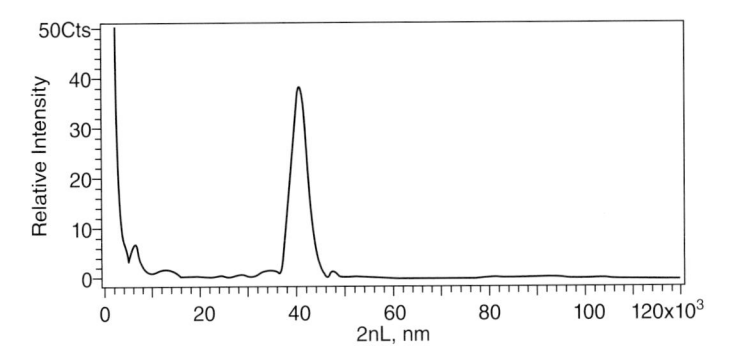

Figure 5.11
Fourier transform of spectrum from Figure 5.10, of a p^{++}-type porous silicon single-layer Fabry–Pérot film, measured in air. This transform was obtained from the region of the reflectance spectrum from 600 to 1000 nm. The value of $2nL$ for this sample is 40 721 nm. Sample is 16 000 nm thick, porosity 67%.

beforehand. A convenient method to analyze these reflectance spectra is from the Fourier transform, which computes the frequency spectrum of an input waveform. An example of a Fourier transform of the reflectance spectrum of the single-layer Fabry–Pérot film from Figure 5.10 is shown in Figure 5.11. The Fourier transform yields a peak whose position along the x-axis corresponds to the fundamental frequency of the cosine term of Equation 5.9, which is equal to the effective optical thickness $2nL$. The Fourier transform of the frequency spectrum of any single or multi-layer porous silicon film will directly yield the product of thickness and refractive index of the various layers in the structure. The method is referred to as RIFTS, for "reflective interferometric Fourier transform spectroscopy" [9, 10]. The following sections go through the technique to obtain and analyze a RIFTS spectrum. We then demonstrate an application of the method to determine the thickness and porosity of a porous silicon layer based on the spectroscopic liquid infiltration method, or SLIM. Additional examples of the application of RIFTS to label-free biosensing can be found in the literature [9–15].

5.3.3.1 Preparation of Spectrum for Fast Fourier Transform

In order to obtain the RIFTS spectrum, the data file must be prepared such that it can be processed by a digital fast Fourier transform (FFT) routine. Obviously the data input to the FFT must be periodic. For Fabry–Pérot layers, this means that the x-axis of the spectrum must be in units of frequency (as in Figure 5.10). A CCD spectrometer usually outputs the spectrum in terms of wavelength, so the x-axis must be inverted. The effect this has on a notional data set is shown in Figure 5.12. Because the original

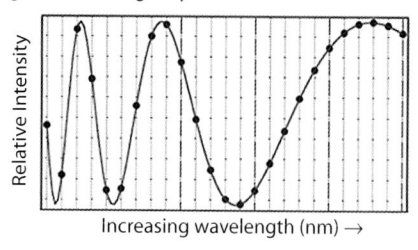

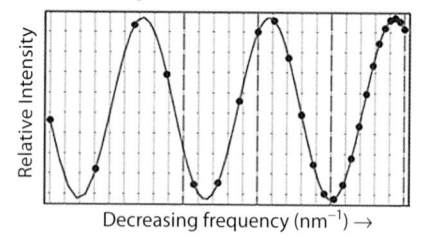

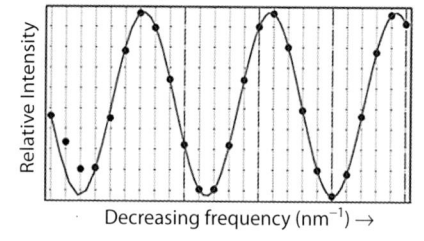

Figure 5.12

Notional data set showing the transformations necessary to prepare an interferometric reflectance spectrum for Fourier transformation. The solid line in each plot represents the ideal Fabry–Pérot interference spectrum, and the solid circles represent the experimental data points that would be output from the CCD spectrometer. (a) The original reflected white light spectrum of the sample, plotted in terms of wavelength. Data points are equally spaced along the x-axis. (b) The wavelength axis is inverted in order for the Fabry–Pérot fringes to be periodic. This causes the point spacing along the x-axis to be nonlinear. (c) The frequency spectrum is interpolated to yield data points that are evenly spaced. Interpolation introduces a systematic error that is apparent in the deviation of the points from the ideal spectrum at higher frequencies.

data points were evenly spaced along the wavelength axis, the inverted axis produces non-linear data point spacing. Most Fourier transform algorithms expect the data points to be evenly spaced along the x-axis, so the x and y positions of those data points have to be determined by interpolation of the spectrum. For this type of data, a cubic spline interpolation generally gives good results. The result of a cubic spline interpolation (based on matching the 2nd derivative of the data) is displayed in the bottom spectrum of Figure 5.12.

5.3.3.2 Interpretation of the Fast Fourier Transform

A rigorous model of the optical properties of porous silicon films should incorporate the frequency dispersion of the refractive index, the effect of multiple reflections, and the instrumental response function. However, the RIFTS method of FFT analysis provides a fast and convenient method of extracting the basic optical parameters, and it yields more reliable data on complicated optical structures, in particular when the optical constants are changing due to analyte admission into the pores during a chemical sensing or biosensing experiment. For most sensing applications, relative changes in these optical constants are as useful as the exact values.

The FFT algorithm has troubles with data sets that truncate suddenly, for example, at the ends of the spectrum. This is remedied by application of a windowing function that smoothly tapers the y-values of the spectrum to zero at either end of the spectrum. A typical windowing function is a Hann (sometimes called Hanning) window, which is a raised cosine function. Once it has been appropriately inverted, interpolated and windowed, the reflectance spectrum is transformed using a digital fast Fourier transform. The real portion of the transform, using the spectrum of Figure 5.10 as an input, is displayed in Figure 5.11. The discrete Fourier transform that generated this spectrum used a multidimensional fast prime factor decomposition algorithm, from the IGOR PRO version 6.0 program library (Wavemetrics, inc, www.wavemetrics.com). A user program that automatically processes Ocean Optics spectral files, written for the IGOR PRO software package, is available at http://sailorgroup.ucsd.edu/software.

There is information contained in both the position and the intensity of a peak in the Fourier transform. As mentioned above, the value of $2nL$ is obtained directly as the position of the peak in the FFT spectrum [16]. The amplitude of the peak is related to the optical constants presented in the Fabry–Pérot interference model of Equation 5.9. The amplitude of the cosine term is related to the amplitude of light reflected at the interfaces, and therefore it is related to the *index contrast* at each pair of interfaces that defines the relevant layer, as given by Equation 5.8 and illustrated in Figure 5.7. The amplitude A_{FFT} of a peak in the FFT spectrum is proportional to this index contrast:

$$A_{FFT} \propto \rho_a \rho_b = \left(\frac{n_{air} - n_{layer}}{n_{air} + n_{layer}} \right) \left(\frac{n_{layer} - n_{Si}}{n_{layer} + n_{Si}} \right) \qquad (5.12)$$

where n_{air}, n_{layer}, and n_{Si} are the refractive indices of air, the porous layer, and bulk silicon, respectively, as defined in Equation 5.8. The relationship of the FFT peak amplitude to the index contrast is only strictly valid if the spectrum from which the FFT derives is an absolute reflectivity spectrum skeleton that makes up the porous material (e.g., Si or SiO$_2$) [8], and n_{layer} is the wavelength-dependent refractive index of the composite porous silicon layer, incorporating both components. The equation can be solved for the refractive index of the composite layer:

$$n_{layer} = \frac{1}{2}\sqrt{\begin{array}{l} 2n_{skeleton}^2 - n_{fill}^2 - 3Pn_{skeleton}^2 + 3Pn_{fill}^2 \\ + \sqrt{8n_{skeleton}^2 n_{fill}^2 + \left(n_{fill}^2 - 2n_{skeleton}^2 + 3Pn_{skeleton}^2 - 3Pn_{fill}^2\right)^2} \end{array}}$$

(5.14)

5.3.4.2 Determination of Thickness and Porosity by SLIM

We can define the experimental observables OT_{air} and OT_{liquid} as:

$$OT_{air} = n_{layer} L \text{ (measured in air)} \tag{5.15}$$

and

$$OT_{liquid} = n_{layer} L \text{ (measured in liquid)} \tag{5.16}$$

Solution of Equation 5.13 for porosity in terms of Equations 5.15 and 5.16 yields:

$$P = 1 - \frac{\left[\left(\frac{OT_{air}}{L}\right)^2 - n_{air}^2\right] \cdot \left[2\left(\frac{OT_{air}}{L}\right)^2 + n_{skeleton}^2\right]}{\left[3\left(\frac{OT_{air}}{L}\right)^2\right] \cdot \left[n_{skeleton}^2 - n_{air}^2\right]} \tag{5.17}$$

and

$$P = 1 - \frac{\left[\left(\frac{OT_{liquid}}{L}\right)^2 - n_{liquid}^2\right] \cdot \left[2\left(\frac{OT_{liquid}}{L}\right)^2 + n_{skeleton}^2\right]}{\left[3\left(\frac{OT_{liquid}}{L}\right)^2\right] \cdot \left[n_{skeleton}^2 - n_{liquid}^2\right]} \tag{5.18}$$

Simultaneous solution of Equations 5.17 and 5.18 yields the values of porosity (P) and thickness (L) for the porous layer. For this analysis it must be assumed that the medium inside the pores (air or liquid) completely fills the porous volume. Computer codes that implement the above calculation are available at http://sailorgroup.ucsd.edu/software. Values of refractive index for some common filling liquids are given in Table 5.3, although it is a good idea to measure the refractive index of the liquid under your laboratory conditions (Mettler Toledo Refracto 30GS refractometer).

5.3.4.3 Determination of Index of Refraction of the Porous Skeleton

The above analysis requires that we have a good idea of the refractive index of the skeleton material. If it has been completely oxidized, the refractive index of the SiO$_2$ skeleton will be close to 1.458, the refractive index of fused

Table 5.3 Index of refraction of some common substances[a].

Compound	Index
Air	1.000272
Methanol, CH_3OH	1.328
Water, H_2O	1.333
Acetone, $CH_3C(O)CH_3$	1.3588
Ethanol, CH_3CH_2OH	1.3611
Hexane, $CH_3(CH_2)_4CH_3$	1.375
Isopropanol, $CH_3CH(OH)CH_3$	1.3776
Dichloromethane, CH_2Cl_2	1.424
Fused silica, SiO_2	1.458
Toluene, $C_6H_5CH_3$	1.497
Quartz, SiO_2	1.544 (n_α), 1.553 (n_β)
Corundum, Al_2O_3	1.761 (n_α), 1.769 (n_β)
Anatase, TiO_2	2.488 (n_α), 2.561 (n_β)
Rutile, TiO_2	2.609 (n_α), 2.900 (n_β)

a) Measured at 20 °C, sodium d line, $\lambda = 589$ nm [24].

silica. This value is fairly constant in the spectral range of interest here (500 to 1000 nm). If the material has not been oxidized, the skeleton can be considered to consist of pure silicon. In that case the refractive index will be wavelength dependent, as shown in Figure 5.9. However, the index in the region 600 to 1000 nm does not deviate much from a value of 3.8, and that value will suffice for most purposes. There are two additional cases where a more complicated analysis is needed: (i) samples that are incompletely oxidized, containing both silicon and silicon oxide in unknown proportions; and (ii) samples containing a third component for which the refractive index is not known–for example, the optical constants of quantum-confined (microporous) silicon can be expected to deviate substantially from those of bulk silicon.

In the first case, for samples containing both elemental silicon and silicon oxide, a detailed fit of the spectrum based on a transfer matrix calculation [4] can provide a precise determination of the volume fractions of the two components, the porosity, and the thickness of the layer [25, 26]. For case (ii), where the optical constants of the layer are unknown, the simplest solution is to assume an average value over the spectral range of interest. This value can be determined in two ways: (i) measurement of the physical thickness (L) of the sample by cross-sectional electron microscopy, and then applying this value to the solutions of Equations 5.16 and 5.17; or (ii) from the relative amplitudes of the FFT spectra using the SLIM method.

Because they measure refractive index contrast at the interfaces of the porous layer, the amplitude of the FFT peaks, A_{FFT}(air) and A_{FFT}(liquid) can provide an estimate of the index of refraction of the porous skeleton. The calculated relationship is shown in Figure 5.14.

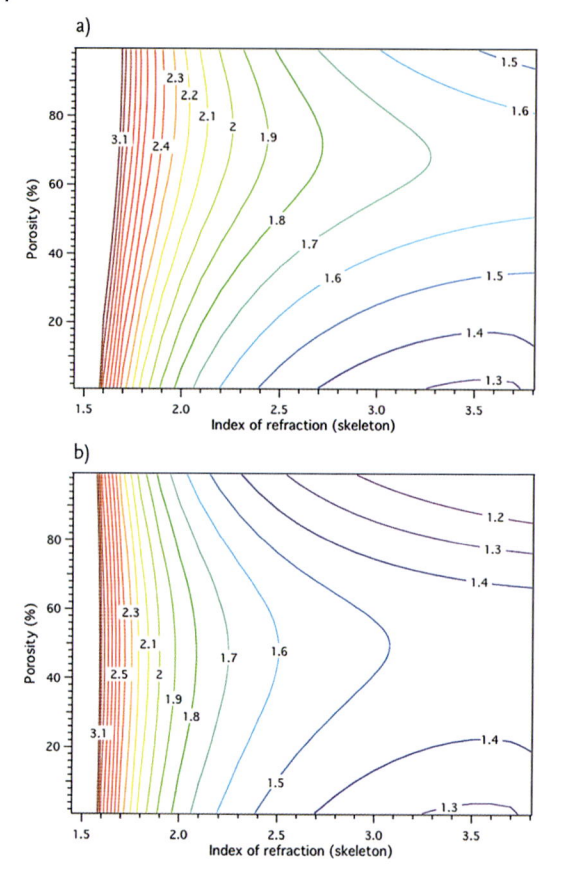

Figure 5.14
Relationship of amplitudes of the two FFT spectra ($A_{FFT(air)}$ and $A_{FFT(liquid)}$, from a SLIM measurement) to porosity and index of refraction of the porous skeleton. Each contour in the plot represents a different value of $A_{FFT(air)}/A_{FFT(liquid)}$ (a), or $(A_{FFT(air)} \cdot OT_{air})/(A_{FFT(liquid)} \cdot OT_{liquid})$ (b); (b) displays less deviation in index for a given contour. If the porosity is known, the index of the porous skeleton can be determined for index values <3. The liquid used in this example is ethanol ($n = 1.3611$ at 20 °C [24]). The curves are determined using the Bruggeman effective medium model.

5.3.4.4 Effect of Skeleton Index on Porosity Determined by SLIM

A nomograph showing the relationship between the two measured OT values (air and liquid ethanol) and porosity is presented in Figure 5.15. Three different values of the refractive index of the skeleton are plotted. The traces were chosen to correspond to a skeleton of pure silicon ($n = 3.8$), a skeleton of pure SiO_2 ($n = 1.455$) and a skeleton of microporous silicon, of notional index 2.1. Note that, for samples whose porosity >75%, the

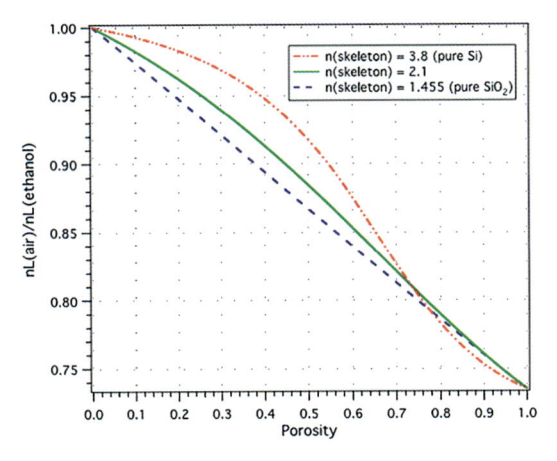

Figure 5.15
Nomograph to determine the porosity of a sample from optical reflectance spectra, based on the spectroscopic liquid infiltration method (SLIM). The curves are determined using the Bruggeman effective medium model, assuming three values for the refractive index of the skeleton of the porous matrix. The three values correspond to a skeleton of pure silicon ($n = 3.8$ at 700 nm), a skeleton of pure SiO_2 ($n = 1.455$ at 700 nm) and a skeleton of microporous silicon, of notional index 2.1. The refractive index of ethanol is taken as 1.3611, its value at 20 °C [24]. To use this nomograph, you first must determine the quantities nL from two reflectance spectra, one of the dry sample (in air) and the other with the sample immersed in pure ethanol. Calculate the ratio $nL_{(air)}/nL_{(ethanol)}$, and read the porosity from the plot. Pure porous SiO_2 can be prepared by air oxidation of porous silicon (see Section 6.1).

index of the skeleton has little influence on the porosity predicted by the ratio $OT_{air}/OT_{ethanol}$. In the porosity range 10 to 70%, the wrong skeleton index can over- or under-predict the true porosity by as much as 20%. Refer to the nomograph of Figure 5.14 to determine the appropriate skeleton index. Automated codes to perform these calculations are available at http://sailorgroup.ucsd.edu/software.

5.3.5
Comparison of Gravimetric Measurement with SLIM for Porosity and Thickness Determination

Both the gravimetric and the SLIM method for determining thickness and porosity are accurate to two or three significant digits [16, 23], and both methods have their limitations. The errors that can be encountered with the gravimetric method were discussed above. Sources of error for the

EXPERIMENT 5.3: Measurement of porosity and thickness of a Fabry–Pérot layer using the Spectroscopic Liquid Infiltration Method (SLIM)

In this experiment you will determine the porosity and thickness of a single-layer porous silicon film from the reflectance spectra measured in air and in ethanol.

Equipment/supplies:

Porous silicon Fabry–Pérot film	Sample for this experiment was prepared following the preparation of "Layer b," one of the Fabry–Pérot test layers prepared for the Bragg stack in Experiment 3.3.
Optical spectrometer and optics	See Table 5.2
Refractometer (optional)	Mettler Toledo Refracto 30GS refractometer. This is so common, you can even get one of these from Amazon (www.amazon.com)

Procedure:

1) Etch a porous silicon chip (1 mΩ cm silicon wafer, 3:1 48% aqueous HF:ethanol, 90 mA cm^{-2}, 300 s). Experiment 3.3 provides a more detailed preparation for a sample derived from p^{++}-silicon. The sample for this experiment was oxidized somewhat, such that it was composed of a mixed silicon/oxide layer.

2) Measure the absolute reflectance spectrum of the sample using a mirror as described in Experiment 5.2.

3) Without moving the sample, wet it with clean, pure (200 proof) ethanol. Record the spectrum using the same conditions as in step (2). Be sure to obtain the spectrum before the ethanol layer becomes too thin. As the ethanol layer begins to dry, it will generate a second Fabry–Pérot layer on top of your sample that may interfere with the reading.

Results:

The spectra and their corresponding Fourier transforms are presented in Figure 5.16. Since this sample is a freshly etched mesoporous/microporous film, the Fourier transform was obtained in the spectral region from 600 to 1000 nm.

The ratio $OT_{air}/OT_{ethanol}$ is 0.83, which sets the value of the porosity between 62 and 70%, depending on the refractive index of the skeleton (Figure 5.15). The ratio $(A_{FFT(air)} \cdot OT_{air})/(A_{FFT(ethanol)} \cdot OT_{ethanol})$ sets the value of the countour for the lower plot of Figure 5.14 at 1.9; this contour yields refractive index values between 1.9 and 2.0 for the porosity range 60 to 70%. For a refractive index of 1.95, the ratio $OT_{air}/OT_{ethanol} = 0.83$ in Figure 5.15 yields a value for porosity ~66%.

Once you have determined the porosity, you can calculate the thickness of your layer. Applying Equation 5.14, the effective refractive index of the layer (n_{layer}) can be calculated from the values of porosity (P), the index of the skeleton material ($n_{skeleton}$), and the index of the material filling the pores (n_{fill}). For this sample in air, $n_{layer} = 1.2789$. Dividing the value of OT_{air} determined from the FFT spectrum (= $nL = 40721/2 = 20361$ nm) by n_{layer} yields $20361/1.2789 = 15920$ nm.

A more precise determination can be obtained by solving Equations 5.16 and 5.17 simultaneously. A numeric approximation (using the SLIM program available at http://sailorgroup.ucsd.edu/software) yields the values porosity = 67.1%, thickness = 15 920 nm, and skeleton index = 1.95.

SLIM method arise from the assumptions associated with the refractive index of the skeleton, optical absorption in the layer, and the existence of voids that are not accessible to the infiltrating liquid. The SLIM method measures the open porosity (pore volume accessible to the probe molecule), while the gravimetric method measures both the open and the closed porosity. Liquid may be excluded from the closed pores for two possible reasons: the pores may be physically closed due to sintering or pore fusion that occurs during processing [25, 26], or the surface tension of the liquid may be too high to adequately wet or infiltrate the smallest pores. Thus SLIM provides a more realistic measure of the accessible volume within the nanostructure. SLIM is a localized measurement (typically a 1-mm spot size of the optical probe), whereas the gravimetric method provides an average over the entire film. The non-destructive nature of SLIM is an advantage for quality control, for example to check reproducibility of an etching procedure. It also provides a means to characterize more complicated optical structures. Experiments 5.4 and 5.5 provide examples of SLIM applied to characterization problems: comparing the extent of oxidation on two porous silicon samples of differing porosity, and measurement of thickness and porosity of a double-layer biosensor.

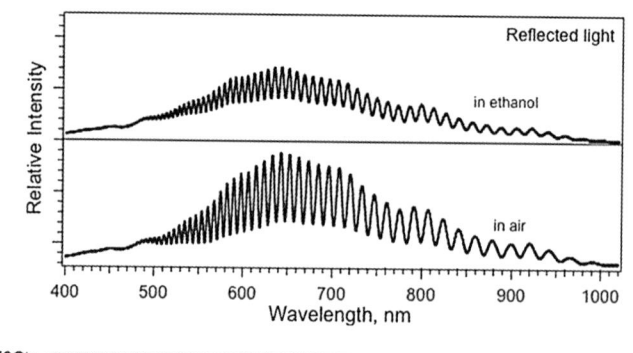

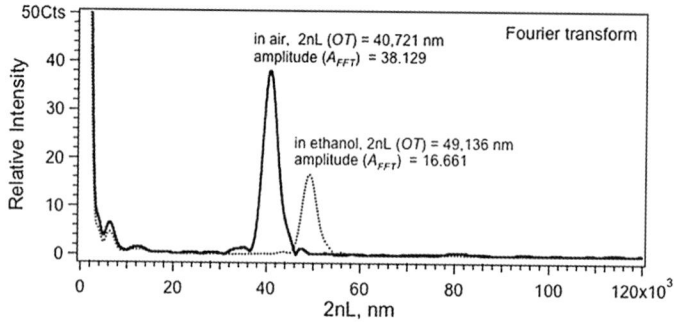

Figure 5.16

Reflected light spectra, obtained in air and in ethanol, and the corresponding Fourier transform spectra needed to determine the porosity and thickness using SLIM. Sample is a p^{++}-type porous silicon single-layer Fabry–Pérot film. The transforms were obtained from the region of the reflectance spectra from 600 to 1000 nm. The values of effective optical thickness ($2nL$) and amplitude (A_{FFT}) of the FFT peak for each spectrum are as indicated.

5.3.6
Analysis of Double-Layer Structures Using RIFTS

As discussed in the previous sections, the value of $2nL$ is obtained directly as the position of the peak in the FFT spectrum. If more than one layer exists in the film, the FFT yields values of $2nL$ for the separate layers as distinct peaks [12]. Thus the interference spectrum for a double-layer structure has an FFT that displays three peaks, corresponding to the values of $2nL$ for layers 1, 2, and 3, respectively, as depicted in Figure 5.18.

Multi-layer porous structures such as these offer some key advantages for biosensing applications. Unlike single-layered Fabry–Pérot films, biosensors based on multilayers can harness optical phenomena other than wavelength shifts. For example, Martin-Palma et al. detected binding of polyclonal mouse antibodies to an amine-modified porous silicon multi-

EXPERIMENT 5.4: Thickness, porosity, and skeleton index for two thermally oxidized porous silicon Fabry–Pérot layers

This experiment illustrates the effect of porous silicon sample porosity on its susceptibility to thermal oxidation. Two samples of different porosity are air-oxidized at 600 °C. They are characterized by SLIM, and the extent of oxidation is determined from the value of the refractive index of the skeleton.

Equipment/supplies:

Two porous silicon Fabry–Pérot films	Samples for this Experiment were prepared in a manner similar to the synthesis of the Fabry–Pérot test layers from Experiment 3.3.
Optical spectrometer and optics	See Table 5.2
Oxidation furnace	Lindberg/Blue M tube furnace or equivalent

Procedure:

1) Etch two porous silicon chips from p^{++}-silicon (1 mΩ cm silicon wafer, 3:1 48% aqueous HF:ethanol, Standard etch cell). Etch "layer 1" at a current density of 500 mA cm^{-2}, for 11 seconds. Etch "layer 2" at a current density of 167 mA cm^{-2}, for 55 seconds. Experiment 3.3 provides more detailed preparation conditions.

2) Insert the samples in the tube furnace and heat to 600 °C. Air-oxidize the samples at this temperature for 60 minutes.

3) After they have cooled, collect optical reflection spectra from both samples, in air and immersed in ethanol, as described in Experiment 5.3. Be careful to obtain both SLIM spectra from the same physical spot on the chip.

Results:

SLIM Data

The peak position and amplitude from the respective FFT spectra (obtained from the raw reflected light spectra, uncorrected for instrumental response) are provided in Table 5.4. As expected from the larger current density used in its preparation, "layer 1" displays a significantly greater porosity than "layer 2." The pore diameters, measured by scanning electron microscopy, are also significantly larger, 50–100 nm versus <20 nm, respectively [12]. The more porous sample is significantly easier to oxidize, and the skeleton index derived from the SLIM analysis reveals that "layer 1" has a refractive index (1.54) close to the value for silica (1.46). Being less porous, the "layer 2" sample displays a larger value for the skeleton index (1.74), indicative of a substantially greater elemental silicon content, and thus a lower extent of oxidation, in this sample. An expansion of the contour plot of Figure 5.14 is shown in Figure 5.17, showing the contours for $A_{FFT(air)}/A_{FFT(ethanol)}$ in the range of $n_{skeleton} = 1.45$ to 2.2. The value of $A_{FFT(air)}/A_{FFT(ethanol)}$ will approach infinity as $n_{skeleton}$ approaches n_{liquid}. This is because the film is index-matched when $n_{skeleton}$ approaches n_{liquid}; the porous silicon/liquid interface is no longer reflective, and the interference fringes vanish. Thus the value of $A_{FFT(air)}/A_{FFT(liquid)}$ increases rapidly for values of $n_{skeleton} < 1.7$. Comparison of the SLIM data to gravimetric measurements of porosity and thickness, and cross-sectional scanning electron microscopy measurements of thickness for these samples are provided in Table 5.5.

Table 5.4 SLIM data for thermally oxidized porous silicon layers.

Porous Si Sample	$2nL$ (nm)	A_{FFT} (counts)	Porosity (%)	Thickness (nm)	Skeleton index
layer 1 (air)	6993.26	255.5	77.1	3140	1.54
layer 1 (ethanol)	8796.57	58.9			
layer 2 (air)	14 142.7	514.9	60.8	5560	1.74
layer 2 (ethanol)	16 750.7	188.9			

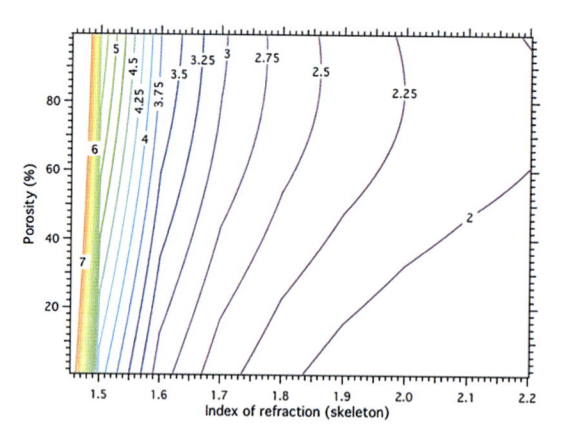

Figure 5.17
Relationship of the amplitudes of the peak in the FFT spectra ($A_{FFT(air)}$ and $A_{FFT(liquid)}$, from a SLIM measurement) to the porosity and index of refraction of the porous skeleton using ethanol as the liquid ($n = 1.3611$ at 20 °C [24]). Each contour in the plot represents a different value of $A_{FFT(air)}/A_{FFT(liquid)}$. The curves are determined using the Bruggeman effective medium model.

5.3 Optical Reflectance Measurements | 165

Table 5.5 Porosity and thickness of thermally oxidized porous silicon layers[a].

Porous Si Layer	Gravimetry		SLIM Measurement		SEM
	porosity (%)	thickness (nm)	porosity (%)	thickness (nm)	thickness (nm)
Layer 1	81	2500	77	3100	2600 ± 400
Layer 2	64	4400	61	5600	5500 ± 200

a) Porosities and thicknesses of porous silicon single layers as determined by gravimetry, by spectral measurement, and by SEM. "layer 1" was etched at 500 mA cm^{-2} for 11 s, and "layer 2" at 167 mA cm^{-2} for 55 s. Both samples were thermally oxidized in air at 600 °C. Data from ([12]).

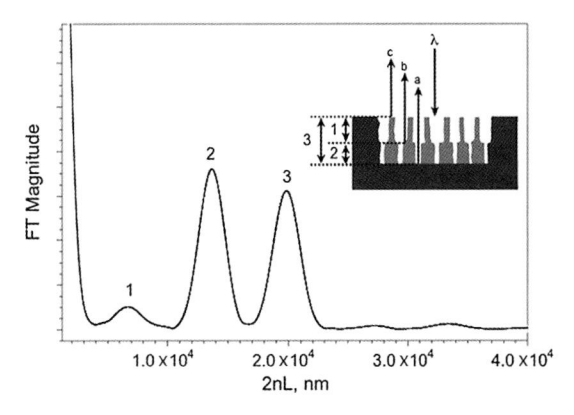

Figure 5.18
Fourier transform of the reflected light spectrum of a double-layer structure, shown schematically in the inset. The number assigned to each peak corresponds to the layer indicated in the inset. Sample is a p^{++}-type porous silicon single-layer Fabry–Pérot film. The transforms were obtained from the region of the reflectance spectrum from 600 to 1000 nm. The values of effective optical thickness ($2nL$) and amplitude (A_{FFT}) of the FFT peak for each spectrum are given in Table 5.6.

layer by observing a reduction in the intensity of the reflected light [27]. Additionally, Chan *et al.* formed porous silicon multilayer structures, such as Bragg mirrors and microcavity resonators, and used modulation of the photoluminescence spectra from these structures to distinguish between Gram(−) and Gram(+) bacteria [28]. Two advantages of using a more elaborate optical structure are that the sensitivity of the measurement can be improved, and drifts caused by thermal fluctuations, changes in sample composition, or degradation of the sensor matrix can be compensated [9–12].

5 Characterization of Porous Silicon

EXPERIMENT 5.5: Thickness, porosity, and skeleton index for a thermally oxidized porous silicon double layer determined by optical reflectance

This experiment analyzes a double-layer structure consisting of a high porosity layer on top of an intermediate porosity layer. It uses the same etching conditions as were used to prepare the two single layers of Experiment 5.4, and it illustrates the ability of RIFTS to determine characteristics of multiple stacked layers simultaneously.

Equipment/supplies:

Porous silicon double layer Fabry–Pérot film	Double layer sample prepared following Experiment 3.1.
Optical spectrometer and optics	See Table 5.2
Oxidation furnace	Lindberg/Blue M tube furnace or equivalent

Procedure:

1) Etch a porous silicon chip from p^{++}-silicon (1 mΩ cm silicon wafer, 3:1 48% aqueous HF:ethanol, Standard etch cell) using the waveform of Figure 3.2, but with the following current-time values: "layer 1" at a current density of 500 mA cm^{-2}, for 11 seconds; "layer 2" at a current density of 167 mA cm^{-2}, for 55 seconds. Experiment 3.1 provides more detailed preparation conditions.

2) Air-oxidize the sample in a tube furnace at 600 °C for 60 minutes. Allow the furnace to cool and remove the sample.

3) Collect SLIM optical reflection spectra from the sample, in air and immersed in ethanol, as described in Experiment 5.3. Be careful to obtain the spectra from the same physical spot on the chip.

Results:

SLIM Data

The peak position and amplitude from the respective FFT spectra (obtained from the raw reflected light spectra, uncorrected for instrumental response) are provided in Table 5.6. For this sample, the FFT peak for "layer 1" in ethanol can be very weak and difficult to identify. In such cases, a second SLIM measurement using another liquid with a significantly different index of refraction, such as methanol (see Table 5.3) could be used. Ideally, the sum of the calculated thickness of "layer 1" and "layer 2" should be equal to "layer 3." Note the close correlation of the thickness, porosity and index values for "layer 1" and "layer 2" in this experiment to the values determined for "layer 1" and "layer 2" in Experiment 5.4; the etch current density and duration for the corresponding layers are the same in both experiments.

Table 5.6 SLIM data for thermally oxidized porous silicon double layer.[a]

Porous Si sample	2nL (nm)	A_{FFT} (counts)	Porosity (%)	Thickness (nm)	Skeleton index
Layer 1 (air)	6705.09	41.0	81	3100	1.45
Layer 1 (ethanol)	8553.03	5.6			
Layer 2 (air)	13 659.10	288.0	63	5460	1.72
Layer 2 (ethanol)	16 281.9	101.8			
Layer 3 (air)	19 855.2	249.0	69	8330	1.67
Layer 3 (ethanol)	24 246.1	81.3			

a) Porous silicon double-layer etched at 500 mA cm^{-2} for 11 s, followed by 167 mA cm^{-2} for 55 s. Sample was thermally oxidized in air at 600 °C for 1 h.

5.4
Porosity, Pore size, and Pore Size Distribution by Nitrogen Adsorption Analysis (BET, BJH, and BdB Methods)

The BET (Brunnauer–Emmett–Teller), BJH (Barret–Joyner–Halenda) and BdB (Broekhof–de Boer) methods are based on the measurement of the adsorption isotherm for weakly interacting gases, such as nitrogen or carbon dioxide at cryogenic temperatures [29–31]. There are commercial instruments that make these measurements, for example, the Micromeritics ASAP 2010 (micromeritics.com). Detailed descriptions of the methods [29, 32] and how they are applied to porous silicon [33] are available in the literature; in this section we provide a simple example of the analysis of two types of porous silicon samples to provide the reader with a basic understanding of the data. Figure 5.19 shows two typical adsorption/desorption isotherms for highly doped p^{++} samples. The pressure (P) of the adsorbate gas is shown along the x-axis, using units relative to the saturation vapor pressure of the gas, P_{sat}. The y-axis represents the volume of gas absorbed by the sample. The adsorption portion of the curve is obtained by increasing the gas pressure to a P/P_{sat} value of unity, and the desorption curve is obtained by reducing P/P_{sat} back to zero, as indicated by the arrows in Figure 5.19.

The samples in Figure 5.19 contain a mixed micro- and meso-porous morphology. The initial portion of the curve, between P/P_{sat} values of 0 and

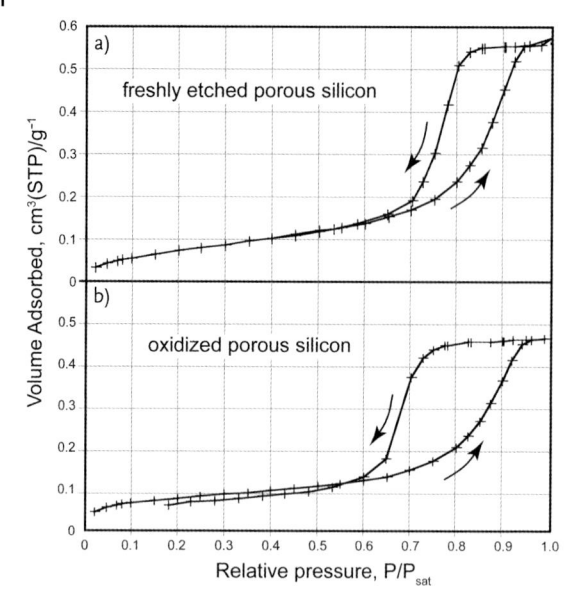

Figure 5.19
Nitrogen adsorption isotherms for a highly doped p^{++}-type porous silicon single-layer. (a) Freshly etched sample, and (b) the same sample type after thermal oxidation (600 °C, 1 h). The curve is typical of a mixed meso- and micro-porous sample. The broader hysteresis loop apparent with the oxidized sample is indicative of smaller pores, and the lower total volume of gas absorbed by this sample indicates a lower total pore volume, relative to the freshly etched sample.

0.1, corresponds to adsorption of a monolayer on the porous surface, and it provides information on the total surface area of the layer (BET method). The strong increase in absorbed volume in the latter portion of the curve (at $P/P_{sat} > 0.6$ for these samples) is indicative of capillary condensation in the nanometer-scale micropores. Finaly, the plateau (observed at P/P_{sat} values > 0.8 for these samples) corresponds to complete filling of the micropores.

The hysteresis loop is due to microcapillary condensation in the smallest pores of the sample. Microcapillary condensation is the physical tendency for a gas to condense in a nanometer-sized pore at temperatures well above the dew point of the gas. The relationship is described by the Kelvin equation [34], which can be used to relate the pore radius to the relative vapor pressure at which condensation occurs:

$$r = -\frac{\gamma V}{RT \ln(P/P_{sat})} \tag{5.19}$$

where r is the radius of the meniscus of the liquid adsorbate, γ is the surface tension of the gas/liquid interface, V is the molar volume of the liquid, R is the gas constant, P_{sat} is the vapor pressure of the adsorbate at temperature T, and P is the observed pressure. The pore radius is larger than the meniscus of the liquid adsorbate by the thickness of the surface adsorbed layer (monolayer or multilayer, but before the pores fill). This model assumes that the pores are fully wetted by the liquid. The smaller the pore radius, the lower the partial pressure at which condensation can occur at a given temperature [35]. Microcapillary condensation is a truly nanoscale phenomenon that provides a means of spontaneously concentrating volatile molecules that is very useful for the sensing of organic vapors [36–49]. Both monolayer adsorption and capillary condensation are influenced by the surface affinity of the porous matrix, which can be tailored by chemical modification [48–50].

Note that the final value of the volume of gas absorbed in the plateau region is less for the oxidized sample than it is for the freshly etched sample. The volume absorbed is directly related to the capacity of the micropores, and the data show how oxidation shrinks the total pore volume in the sample. The simple interpretation is that oxidation adds two oxygen atoms for every atom of silicon in the porous skeleton, and the pore walls swell as a consequence of this volume increase. Swelling of the pore walls reduces the volume of the pores. It also reduces the average pore diameter in the sample, which is indicated by the hysteresis loop.

Micropore diameter affects the shape of the hysteresis loop observed during an adsorption/desorption cycle. This hysteresis derives from the shape of the pores, and from the fact that the condensation and evaporation processes are not the exact reverse of each other. The simplified situation is described pictorially in Figure 5.20. During adsorption, the pore fills by adding monolayers to the entire pore surface, and so the growing liquid

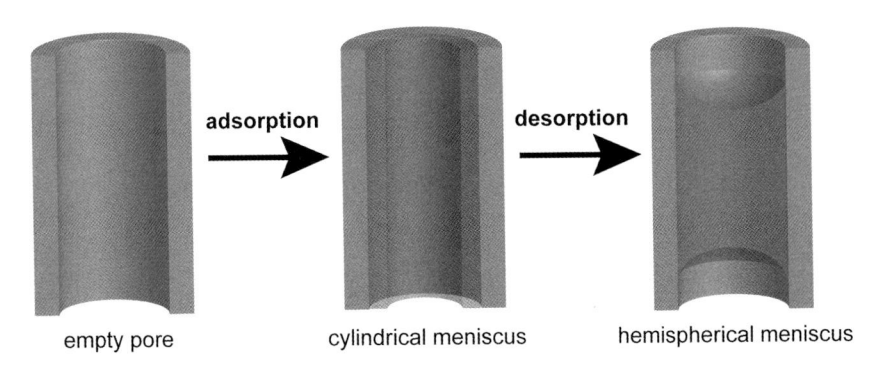

Figure 5.20
Adsorption and desorption in a cylindrical micropore. The processes follow two separate pathways, which leads to the hysteresis observed in the gas sorption isotherms.

front is cylindrical in shape. The desorption process removes adsorbate, starting from the outside and moving into the pore. The receding liquid interface is hemispherical in shape. The asymmetry of these two processes generates the hysteresis observed, which is highly dependent on the surface tension of the liquid and the diameter of the pore. The width of the hysteresis loop provides a measure of this pore diameter. Fitting of these curves to idealized pore (cylindrical) and adsorbate interaction models is the basis of the BJH and the BdB methods.

Nitrogen adsorption methods are especially sensitive to small pore structures, in particular micropores (radii below 1 nm) [51] that are not easily observable in SEM micrographs. The micropores are emphasized in the BET measurement, which is relatively insensitive to macropores. For large mesopore and macropore structures, the changes in relative pressure might be too small to allow reliable determination of the average pore size using the BET approach, and electron microscopy can be a better means to determine average pore size. Be aware that the SEM can be deceiving: what might appear to be solid silicon walls in a "macroporous" silicon sample can actually be a microporous sponge. Thus, applying both electron microscopy *and* the BET method will provide a more complete picture of the pore morphology. The BET method also provides the best means to measure the specific surface area of a porous sample.

5.5
Measurement of Steady-State Photoluminescence Spectra

5.5.1
Origin of Photoluminescence from Porous Silicon

In 1960, Allen Gee at Texas Instruments reported that driving a constant anodic current through a stain-etched porous silicon sample results in a "faint, reddish glow." No mechanism was proposed, and the paper concludes with the statement ". . . Further efforts are directed toward locating the region within the film from which light is emitted and toward elucidating the mechanisms involved". Then, in the late 1980s, two groups independently and simultaneously recognized that the diaphanous filaments present in high porosity porous silicon might display quantum confinement effects [52, 53]. Volker Lehmann and Ulrich Goesele reported that electrochemical formation of porous silicon was intimately tied to the generation of nanometer-scale quantum wires [53], and they showed that the optical absorption spectrum was consistent with a quantum confinement model. Leigh Canham and his coworkers then showed that the material emits a bright red–orange photoluminescence, a clear indication that the optical physics of the nanomaterial differs from bulk crystalline silicon [52].

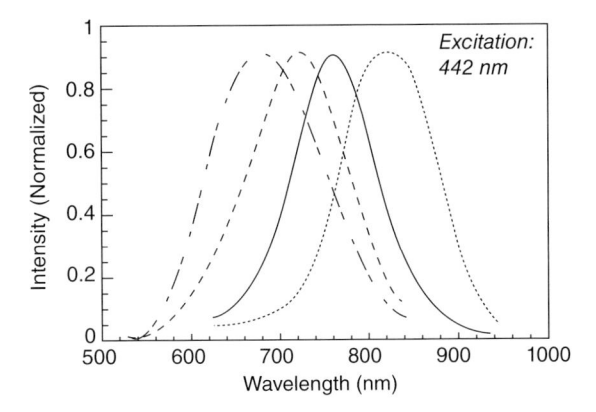

Figure 5.21
Photoluminescence spectra of four n-type porous silicon samples, prepared by etching at different current densities. The photoluminescence spectrum is dependent on preparation conditions – in particular the current density. From reference [56].

Photoluminescence in the visible region of the spectrum corresponds to energies significantly larger than the bandgap energy of silicon (1.12 eV, or 1100 nm).

5.5.1.1 Tunability of the Photoluminescence Spectrum

In his first paper describing visible photoluminescence from porous silicon, Canham reported that the spectrum of light given off from the material could be tuned from near-infrared to green by adjusting the preparation conditions [52]. For example, Figure 5.21 shows a set of photoluminescence spectra obtained from samples prepared under slightly different etching current densities. Canham showed that the emission spectrum from porous silicon can be shifted to the blue by soaking the sample in a hydrofluoric acid solution [52]. This result was interpreted as a manifestation of quantum confinement effects; as a nanocrystal gets smaller its band gap should increase, and its emission spectrum shift to the blue [54, 55].

5.5.1.2 Mechanisms of Photoluminescence

Despite its origin in the nanometer-scale structures of a porous silicon matrix, there is not a great deal of correlation between the gross morphology of a porous silicon layer and its photoluminescence spectrum. Samples that appear macroporous, mesoporous, or microporous by SEM can all yield photoluminescence signals. This is, in part, due to the fact that all types of porous silicon morphologies can contain nanometer-scale features [55–60]. However, other emission mechanisms can be involved. With the

Table 5.7 Luminescence bands associated with porous silicon[a].

Spectral range	Peak wavelength (nm)	Luminescence band label
Ultraviolet	~350	UV band
Blue–green	~470	F band
Green–red	500–800	S band
Near-infrared	1100–1500	IR band

a) Adapted from reference ([61]).

correct preparation, porous silicon can be coerced to emit all the way into the blue end of the visible spectrum (420 nm) [58, 59], although in this case the emission mechanism is related to oxide-based species rather than simple quantum confinement [59]. It is now generally agreed that photoluminescence from porous silicon can derive from a combination of quantum confinement and surface effects [60]. The relative contribution of quantum confinement, defects, interfacial oxides, hydrides, and chemical impurities to the photoluminescence spectrum has led to a lively and sometimes confusing dialog in the scientific literature regarding the mechanism of emission from porous silicon [55, 60–68]. The four main types of photoluminescence phenomena observed from porous silicon samples are summarized in Table 5.7.

Even if they contain nanometer scale silicon features, not all porous silicon samples are photoluminescent. A porous silicon sample prepared from n-type silicon (like those prepared in Experiments 2.3 or 2.4) will generally display orange photoluminescence right out of the etch bath. The light emission in this case comes primarily from quantum confinement in nanocrystalline silicon domains formed during the pore etching process [69], and it is designated the S band (Table 5.7). The "S" label stands for "slow," because the emission lifetime for this material is on a timescale of microseconds. By contrast, the "fast" F band decays in a few nanoseconds. The UV and IR band designations are self-explanatory, labeled for the region of the spectrum in which they appear. The UV and F bands are thought to arise from oxide-based defects, as they are only seen with oxidized material. The IR band has been ascribed to "dangling bond" states [61], or a silicon atom that is only bonded to three other atoms rather than its usual complement of four.

To display bright S band photoluminescence, samples must contain a large number of quantum-sized silicon crystallites, and these cannot be quenched by surface species or defects. The etching conditions, and in many cases the post-etching treatments, are crucial in determining the intensity and emission maximum of photoluminescence. Because photoluminescence is so sensitive to the surface, the intentional introduction of chemical species at the surface that can open up a quenching pathway is a useful means to make a sensor from porous silicon [70].

5.5.2
Instrumentation to Acquire Steady-State Photoluminescence Spectra

Light given off in the form of photoluminescence tends to be much weaker than reflected light. However, the CCD spectrometer that we used to collect the reflected light spectra in Experiments 5.2 through 5.5 can also detect photoluminescence, provided the photoluminescence quantum yield of the sample is sufficient. The as-etched n-type porous silicon sample prepared in Experiment 2.3 will have an external quantum yield of between 1 and 10% when excited with a UV (380 nm) source. This can be detected with the sensor in an Ocean Optics USB-4000 spectrometer. Weaker spectra require a cooled detector, such as a Princeton Instruments Pixis or SPEC-10 (www.princetoninstruments.com/). Ocean Optics also sells a thermoelectrically cooled detector/spectrometer package in their QE-65000.

An adequate photoluminescence measurement system can be put together with slight modification of the optical design used to collect reflectance data (Figure 5.4). The tungsten light source must be replaced with a blue or UV source. LED sources, such as the 365 nm Ocean Optics LLS-365, provide a low-cost and stable solution. If you take this route, you also need to place a cut-off filter in the beam path going to the detector to eliminate the 365 nm line and its overtones from the spectrum. That will require a filter holder (Ocean Optics FHS-UV In-Line Filter Holder) and a filter (Newport Corporation 10LWF-400-B, www.newport.com) that absorbs the UV light but passes the visible and near-IR spectrum (referred to as a long-pass filter). The bifurcated optical fiber employed in Figure 5.4 is not ideal for fluorescence measurements. To avoid interfering fluorescence signals from the optical components, it is better to separate the excitation source from the detection optics using two optical fibers. A recommended set-up is shown in Figure 5.22.

5.6
Time-Resolved Photoluminescence Spectra

5.6.1
Long, Nonexponential Excited State Lifetimes

Unlike many molecular fluorophores, porous silicon has a relatively long-lived excited state that decays in a nonexponential fashion. From the method

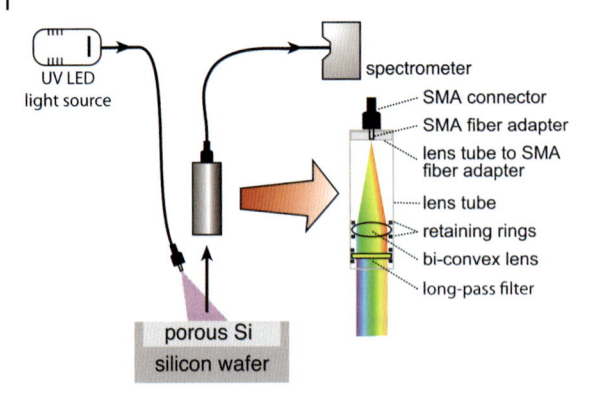

Figure 5.22

Schematic diagram of the optical set-up used to collect fluorescence spectra. A UV (or blue) LED light source is focused onto the sample surface by means of an optical fiber (Ocean Optics QP600-1-UV-VIS 600 μm Premium Fiber, UV/VIS, 1 m, www.oceanoptics.com). A CCD spectrometer is coupled to a microscope objective lens using a second fiber optic cable. The light collection lens assembly contains a long-pass filter (Newport Corporation 10LWF-400-B, www.newport.com) to eliminate the blue or UV excitation light, and the fluorescence light is coupled into the optical fiber that then goes to the spectrometer. The other parts are listed in Table 5.2.

of its synthesis, it is easy to imagine how porous silicon could consist of an ensemble of quantum structures of varying sizes, giving rise to the broad emission spectra seen in Figure 5.21 (FWHM of 100–150 nm) [71]. Time-resolved spectra (Figure 5.23) indicate that the ensemble displays a distribution of emission lifetimes [72]. The excited states with shorter emission wavelengths decay more quickly, with half-lives of the order of several tens of microseconds at the red end of the spectrum, and less than 5 μs at the blue end [73–75]. The photoluminescence also exhibits an interesting polarization memory; if the excitation source is polarized, photoemission can also be polarized [76, 77]. To complicate matters further, energy exchange can occur between the quantum-confined structures in the porous silicon matrix [78–81].

The photoluminescence decay curves in Figure 5.23 are nonexponential, and this is typical of porous silicon samples. Various approaches to obtain meaningful lifetime information from this type of data have been attempted, including the use of a stretched exponential fit [82, 83] and calculation of an "effective τ" based on integration of the intensity versus time curve [84, 85]. A fit to the empirical Kohlraush–Williams–Watts stretched exponential model is shown in Figure 5.23. The stretched exponential is often used to

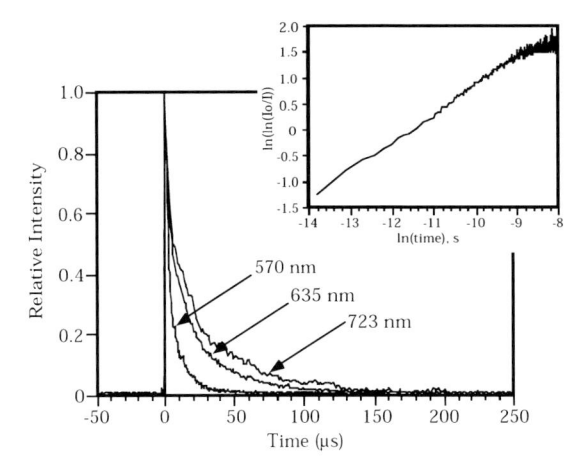

Figure 5.23
Photoluminescence decays from a single porous silicon sample, obtained at three different emission wavelengths, as indicated. Porous silicon samples typically display a longer lifetime at the red end of the spectrum. Inset shows Kohlrausch–Williams–Watts stretched exponential fit of the decay measured at 635 nm. From reference [56].

describe systems that exhibit a distribution of emission lifetimes [86]. Decay rate constants obtained in this fashion are not particularly rigorous, but they allow relative comparisons to be made [70].

5.6.2
Influence of Surface Traps

Nonradiative carrier recombination (i.e., recombination that does not result in emission of a photon) can occur at defects on semiconductor surfaces. This is a particular problem for gallium arsenide and many of the II–VI semiconductors, which do not form electronically passive native oxides. One of the reasons for the use of silicon in microelectronic circuits is that the oxides and hydrides of silicon make very passive interfaces. With the large surface-to-volume ratio present in porous silicon, the effect of surface defects on electronic properties is even more pronounced [60]. For a nanometer-sized feature, up to one third of the atoms can reside at the surface, and these surface atoms can exert a pronounced influence on the emission spectrum.

The electronic states associated with the "core" of a semiconductor nanoparticle can be in communication with surface states [87, 88], and this is definitely true for porous silicon. Koch has referred to this as the "smart" quantum confinement model [68], and asserts that there is extensive coupling between the quantum confined states in the core and the surface wavefunction in porous silicon. The surface species responsible may be Si–H_x, Si–OH, and Si–O–Si moieties silicon dangling bonds, or lattice defect sites. These species can introduce intra-bandgap states that allow sub-bandgap emission or nonradiative relaxation to occur [89–91].

Researchers have harnessed various chemical reactions in order to either (i) passivate these suface states to improve photoluminescence output [59, 92–96], or (ii) capitalize on the surface sensitivity for specific chemical sensor applications [70]. Chapter 6 will cover the general chemical transformations available to the porous silicon system. The next section discusses one of the main tools used to characterize these surface species, infrared spectroscopy.

5.7
Infrared Spectroscopy of Porous Silicon

5.7.1
Characteristic Group Frequencies for Porous Silicon

Fourier-transform infrared (FTIR) provides the most convenient method for characterization of chemical species on porous silicon surfaces. Absorbances in the mid-infrared spectrum (frequencies of 4000 to 200 cm^{-1}, corresponding to the wavelength range 2500–50000 nm) arise primarily from molecular vibrations, and they are characteristic of the types of atoms and chemical bonds in the sample. Chemists refer to this as the "fingerprint" region of the infrared spectrum. The high surface area of porous silicon places a large quantity of the material in the beam path, resulting in a high signal-to-noise ratio, even for species that are present at relatively low fractional surface coverage; the FTIR signals from porous silicon are typically 1000 times stronger compared with the vibrational spectrum of a flat silicon surface. Table 5.8 lists characteristic band frequencies associated with porous silicon and some common surface species, along with their assignments.

Raman spectroscopy is a vibrational characterization technique that is complementary to the infrared method. The selection rules for Raman activity are different than those for infrared; in particular, totally symmetric vibrations are allowed in Raman, whereas they are forbidden in infrared. This makes Raman a good means to characterize silicon lattice vibrations, which are more difficult to measure by FTIR [103–106].

Table 5.8 Common infrared bands associated with porous silicon.

Species	Frequency (cm^{-1})[a]	Mode assignment	Reference	Comment
Si–CH$_3$	2965	ν_a C–H	[97]	Ref. includes D isotope data
Si–CH$_3$	2898	ν_s C–H	[97]	Ref. includes D isotope data
O$_3$SiH	2268	ν Si–H	[98, 99]	Ref. [98] includes D isotope data. Ref. [99] places this band at 2246 cm^{-1} (wet air oxidized) or 2256 cm^{-1} (dry air oxidized)
O$_2$SiH$_2$	2200	ν Si–H	[98, 99]	Ref. [98] includes D isotope data. Ref. [99] assigns band for wet air oxidized sample at 2192 cm^{-1}
SiH$_3$	2140	ν Si–H	[100]	
OSiH$_3$	2160	ν Si–H	[98, 99]	Ref. [98] includes D isotope data, places this band at 2119 cm^{-1}
SiH$_2$	2108	ν Si–H	[100]	
Si–H	2087	ν Si–H	[98, 100]	Ref. [98] includes D isotope data. Assignment from Ref. [100]
Si–CH$_3$	1400	δ_a CH$_3$	[97]	Ref. includes ^{13}C isotope data
Si–CH$_3$	1252	δ_s CH$_3$	[97]	"umbrella mode" Ref. includes ^{13}C isotope data
Si–O–Si	1150–1240	ν_s Si–O–Si	[101, 102]	Similar to LO mode of crystalline SiO$_2$ (1240 cm^{-1}) or A$_2$ mode of fused silica (1190 cm^{-1}). Exact position dependent on thermal history of sample. Usually weaker than ν_a Si–O–Si mode.
Si–O–Si	980–1050	ν_a Si–O–Si	[98, 99, 101, 102]	Very strong intensity band. Similar to TO mode of crystalline SiO$_2$ (1050 cm^{-1}) or E mode of fused silica (1110 cm^{-1}). Exact position dependent on thermal history of sample

(*Continued*)

5 Characterization of Porous Silicon

Table 5.8 (Continued)

Species	Frequency (cm^{-1})[a]	Mode assignment	Reference	Comment
SiH$_2$	913	δ_s Si–H$_2$	[98, 100]	Medium intensity band. "scissors mode" Ref. [98] includes D isotope data
O$_3$Si–H	857	δ Si–H	[98]	
O$_2$Si–H	820	δ Si–H	[98]	
Si–F	812	ν Si–F	[100]	Weak band. Etched Si always has trace quantities of F.
OSi–H	775	δ Si–H	[98]	
Si–CH$_3$	766	ρ Si–CH$_3$	[97]	Strong band. The most characteristic band associated with Si–CH$_3$, on the surface, it is not seen with hydrocarbons other than CH$_3$
Si–CH$_3$	681	ν Si–C	[97]	Weak band, usually obscured by other bands. Assignment substantiated by ^{13}C and D data
Si–H$_x$	665	δ Si–H$_2$	[98, 100]	Strong band. "Wagging" vibration; highly coupled with other modes [100]
Si–H$_x$	628	δ Si–H	[98, 100]	Very Strong band. Highly coupled with other modes [100]
Si–Si	610	ν Si–Si	[100]	Weak, Si lattice 2-phonon TO+TA mode

a) All values ± 5 cm^{-1}.
ν = stretching mode; δ = deformation mode; ρ = rocking mode; a = antisymmetric; s = symmetric.

5.7.2
Measurement of FTIR Spectra of Porous Silicon

One of the main challenges involved in acquisition of FTIR spectra of porous silicon samples is that free carriers in silicon strongly absorb infrared radiation. The problem is most pronounced with highly doped silicon – in particular when using a transmission configuration where the infrared beam must pass through the substrate along with the porous silicon layer. Silicon with resistivity $> 1\,\Omega\,cm$ can be measured in transmission mode with a laboratory-grade spectrometer such as the Thermo Scientific Nicolet 6700 FTIR (www.thermoscientific.com), but samples with resistivity $<0.1\,\Omega\,cm$ are typically too strongly absorbing to provide reasonable signals in transmission mode. These latter samples are more conveniently measured using a diffuse reflectance or attenuated total reflectance (ATR) configuration, where the probing radiation does not need to pass through the 300 to 500 µm thick silicon substrate. Diffuse reflectance and ATR attachments that are inserted into the sample chamber of the optical bench are commercially available for virtually all of the research-grade FTIR instruments on the market. Typical ATR-FTIR spectra of freshly etched and thermally oxidized porous silicon samples are shown in Figure 5.24.

A second problem encountered with FTIR measurements is Fabry–Pérot interference. This appears in the spectrum as a rolling, or sinusoidal baseline, and arises from constructive and destructive interference of the infrared beam in the porous silicon layer. While optical Fabry–Pérot interference can provide a powerful means to characterize porous Si films (described earlier in this chapter), in infrared spectra it is more of an annoyance that obscures the peaks of interest. Interference fringes are more pronounced with smooth films, and they are most pronounced in transmission mode spectra. Due to its optical configuration, Fabry–Pérot interference is not observed in ATR spectra.

5.7.2.1 Transmission Mode Measurement Using the Standard Etch Cell

The Standard etch cell (Appendix 1) has slots machined in the base that allow direct mounting of a sample into the FTIR instrument, allowing reproducible acquisition of spectra from the same spot on the sample, which is convenient, for example, for following the course of a chemical modification reaction. This will only work for lightly doped (n- or p-type) silicon, which is sufficiently transparent in the mid-IR spectral region. In transmission mode, the FTIR absorbance values are directly proportional to the concentration of the species generating the characteristic vibrations. A transmission-mode geometry will thus yield quantitative absorbance data for many species of interest. The infrared absorptions of Si–O stretching vibrations have a high extinction coefficient, providing excellent sensitivity if one wishes to probe the degree of oxidation of a porous silicon sample. The Si–H stretching modes are also very sensitive to the degree of oxidation of the sample; when one or more oxygen atoms is bound to a surface silicon atom that also contains one or more hydrogen atoms, the Si–H vibrations shift to higher frequency (Table 5.8 and Figure 5.24).

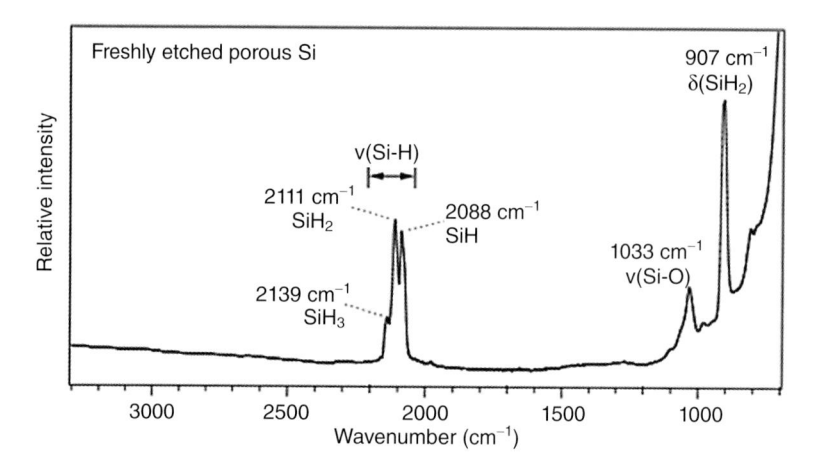

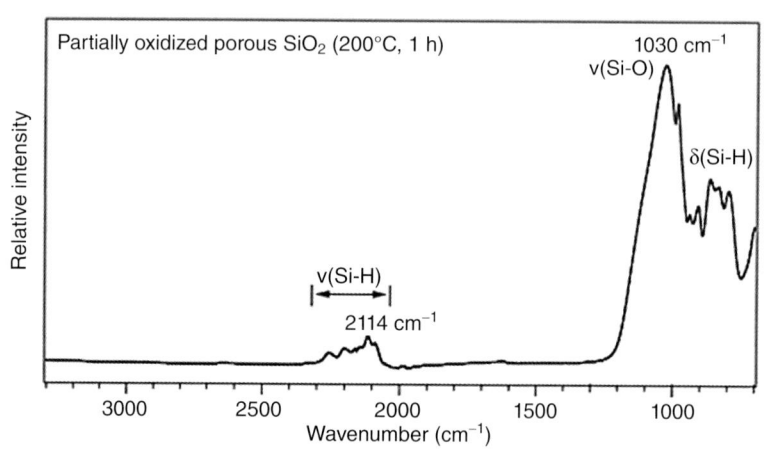

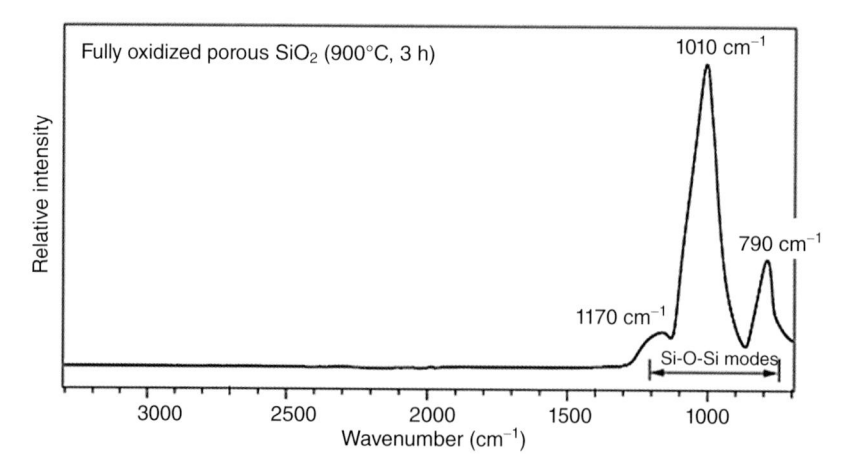

5 Characterization of Porous Silicon

Freshly etched porous Si

- 2111 cm^{-1} SiH$_2$
- 2088 cm^{-1} SiH
- 2139 cm^{-1} SiH$_3$
- ν(Si–H)
- 1033 cm^{-1} ν(Si–O)
- 907 cm^{-1} δ(SiH$_2$)

Partially oxidized porous SiO$_2$ (200°C, 1 h)

- ν(Si–H) 2114 cm^{-1}
- 1030 cm^{-1} ν(Si–O)
- δ(Si–H)

Fully oxidized porous SiO$_2$ (900°C, 3 h)

- 1010 cm^{-1}
- 1170 cm^{-1}
- 790 cm^{-1}
- Si–O–Si modes

Figure 5.24
ATR-FTIR spectra of freshly etched and thermally oxidized porous silicon samples, obtained with a Thermo Scientific Nicolet Smart iTR diamond ATR attachment on a Thermo Scientific Nicolet 6700 FTIR instrument. Common vibrational assignments are indicated. The spectrum of the "fully oxidized" sample displays bands consistent with the main bands of fused silica, at 800 cm^{-1} (m), 950 cm^{-1} (w), 1100 cm^{-1} (vs), and 1190 cm^{-1} (m) [102]. These spectra were obtained from a p-type porous silicon sample, etched in 3:2 aqueous 48% HF:ethanol, at a current density of 20 mA cm^{-2} for 20 min. The porosity and thickness of the freshly etched porous layer were 50% and 2.6 μm, respectively. The porosity of the "fully oxidized" sample is significantly lower (15%), and the thickness of the layer is significantly smaller (1.5 μm) than the other two samples, indicating that the pores have begun to collapse at this high temperature.

References

1. Halimaoui, A. (1997) Porous silicon formation by anodisation, in *Properties of Porous Silicon*, vol. 18 (ed. L. Canham), Institution of Engineering and Technology, London, pp. 12–22.
2. Segal, E., Perelman, L.A., Cunin, F., Renzo, F.D., Devoisselle, J.-M., Li, Y.Y., and Sailor, M.J. (2007) Confinement of thermoresponsive hydrogels in nanostructured porous silicon dioxide templates. *Adv. Funct. Mater.*, **17**, 1153–1162.
3. Hecht, E. (1998) *Optics*, 3th edn, Addison-Wesley, Reading, MA, pp. 377–428.
4. Born, M., and Wolf, E. (1999) *Principles of Optics: Electromagnetic Theory of Propagation, Interference and Diffraction of Light*, 7th edn, Cambridge University Press, New York, p. 952.
5. Macleod, H.A. (1969) *Thin-Film Optical Filters*, Adam Hilger LTD, London.
6. Bruggeman, D.A.G. (1935) Berechnung verschiedener physikalischer konstanten von heterogenen substanzen. *Ann. Phys.*, **24**, 636–679.
7. Looyenga, H. (1965) Dielectric constants of heterogeneous mixtures. *Physica*, **31**, 401–406.
8. Palik, E.D., and Holm, R.T. (1998) *Handbook of Optical Constants of Solids*, Academic Press, New York.
9. Wu, J., and Sailor, M.J. (2009) Chitosan hydrogel-capped porous SiO_2 as a pH-responsive nano-valve for triggered release of insulin. *Adv. Funct. Mater.*, **19**, 733–741.
10. Pacholski, C., Perelman, L.A., VanNieuwenhze, M.S., and Sailor, M.J. (2009) Small molecule detection by reflective interferometric Fourier transform spectroscopy (RIFTS). *Phys. Status Solidi A*, **206** (6), 1318–1321.
11. Pacholski, C., Yu, C., Miskelly, G.M., Godin, D., and Sailor, M.J. (2006) Reflective interferometric fourier transform spectroscopy: a self-compensating label-free immunosensor using double-layers of porous SiO_2. *J. Am. Chem. Soc.*, **128**, 4250–4252.
12. Pacholski, C., Sartor, M., Sailor, M.J., Cunin, F., and Miskelly, G.M. (2005) Biosensing using porous silicon double-layer interferometers: reflective interferometric Fourier transform spectroscopy. *J. Am. Chem. Soc.*, **127** (33), 11636–11645.
13. Mun, K.-S., Alvarez, S.D., Choi, W.-Y., and Sailor, M.J. (2010) A stable, label-free optical

50 Gao, T., Gao, J., and Sailor, M.J. (2002) Tuning the response and stability of thin film mesoporous silicon vapor sensors by surface modification. *Langmuir*, **18** (25), 9953–9957.

51 Hérino, R. (1997) Pore size distribution in porous silicon, in *Properties of Porous Silicon*, vol. 18 (ed. L. Canham), Institution of Engineering and Technology, London, pp. 89–96.

52 Canham, L.T. (1990) Silicon quantum wire array fabrication by electrochemical and chemical dissolution. *Appl. Phys. Lett.*, **57** (10), 1046–1048.

53 Lehmann, V., and Gosele, U. (1991) Porous silicon formation: a quantum wire effect. *Appl. Phys. Lett.*, **58** (8), 856–858.

54 Brus, L. (1987) Size dependent development of band structure in semiconductor crystallites. *Nouv. J. Chim.*, **11** (2), 123.

55 Collins, R.T., Fauchet, P.M., and Tischler, M.A. (1997) Porous silicon: from luminescence to LEDs. *Phys. Today*, **50**, 24–31.

56 Sailor, M.J., Heinrich, J.L., and Lauerhaas, J.M. (1997) Luminescent porous silicon: synthesis, chemistry, and applications, in *Semiconductor Nanoclusters: Physical, Chemical, and Catalytic Aspects*, vol. 103 (eds P.V. Kamat and D. Meisel), Elsevier Science B. V., Amsterdam, pp. 209–235.

57 Doan, V.V., Penner, R.M., and Sailor, M.J. (1993) Enhanced photoemission from short-wavelength photochemically etched porous silicon. *J. Phys. Chem.*, **97**, 4505–4508.

58 Gelloz, B., Mentek, R., and Koshida, N. (2009) Specific blue light emission from nanocrystalline porous Si treated by high-pressure water vapor annealing. *Jpn. J. Appl. Phys.*, **48** (4), 04c119.

59 Gelloz, B., and Koshida, N. (2009) Long-lived blue phosphorescence of oxidized and annealed nanocrystalline silicon. *Appl. Phys. Lett.*, **94** (20), 201903.

60 Sa'ar, A. (2009) Photoluminescence from silicon nanostructures: the mutual role of quantum confinement and surface chemistry. *J. Nanophotonics*, **3**, 032501.

61 Cullis, A.G., Canham, L.T., and Calcott, P.D.J. (1997) The structural and luminescence properties of porous silicon. *J. Appl. Phys.*, **82** (3), 909–965.

62 Mizuno, H., Koyama, H., and Koshida, N. (1996) Oxide-free blue photoluminescence from photochemically etched porous silicon. *Appl. Phys. Lett.*, **69** (25), 3779–3781.

63 Prokes, S.M., and Glembocki, O.J. (1995) Role of interfacial oxide-related defects in the red-light emission in porous silicon–reply. *Phys. Rev. B*, **51** (16), 11183–11186.

64 Banerjee, S. (1995) Role of interfacial oxide-related defects in the red-light emission in porous silicon–comment. *Phys. Rev. B*, **51** (16), 11180–11182.

65 Banerjee, S., Narasimhan, K.L., and Sardesai, A. (1994) Role of hydrogen- and oxygen- terminated surfaces in the luminescence of porous silicon. *Phys. Rev. B*, **49** (4), 2915–2918.

66 Searson, P.C., Prokes, S.M., and Glembocki, O.J. (1993) Luminescence at the porous silicon/electrolyte interface. *J. Electrochem. Soc.*, **140** (11), 3327–3331.

67 Qin, G.G., and Jia, Y.Q. (1993) Mechanism of the visible luminescence in porous silicon. *Solid State Comm.*, **86** (9), 559–563.

68 Koch, F., Petrova-Koch, V., and Muschik, T. (1993) The luminescence of Porous Silicon: the case for the surface state mechanism. *J. Lumin.*, **57**, 271–281.

69 Nash, K.J., Calcott, P.D.J., Canham, L.T., and Kane, M.J. (1994) The

origin of efficient luminescence in highly porous silicon. *J. Lumin.*, **60–61**, 297–301.
70. Sailor, M.J., and Wu, E.C. (2009) Photoluminescence-based sensing with porous silicon films, microparticles, and nanoparticles. *Adv. Funct. Mater.*, **19** (20), 3195–3208.
71. Yorikawa, H., and Muramatsu, S. (1998) Analysis of photoluminescence from porous silicon: particle size distribution. *J. Appl. Phys.*, **84** (6), 3354–3358.
72. Bsiesy, A., Vial, J.C., Gaspard, F., Hérino, R., Ligeon, M., Muller, F., Romestain, R., Wasiela, A., Halimaoui, A., and Bomchil, G. (1991) Photoluminescence of high porosity and of electrochemically oxidized porous Si layers. *Surf. Sci.*, **254**, 195–200.
73. Andrianov, A.V., Kovalev, D.I., Shuman, V.B., and Yaroshetskii, I.D. (1992) Short lived green band and time evolution of the photoluminescence spectrum of porous silicon. *JETP Lett.*, **56** (5), 236–239.
74. Mihalcescu, I., Vial, J.C., and Romestain, R. (1996) Carrier localization in porous silicon investigated by time resolved luminescence analysis. *J. Appl. Phys.*, **80** (4), 2404–2411.
75. Xie, Y.H., Wilson, W.L., Ross, F.M., Mucha, J.A., Fitzgerald, E.A., Macaulay, J.M., and Harris, T.D. (1992) Luminescence and structural study of porous silicon films. *J. Appl. Phys.*, **71** (5), 2403–2407.
76. Efros, A.L., Rosen, M., Averboukh, B., Kovalev, D., Ben-Chorin, M., and Koch, F. (1997) Nonlinear optical effects in porous silicon: photoluminescence saturation and optically induced polarization anisotropy. *Phys. Rev. B*, **56** (7), 3875–3884.
77. Kovalev, D., Ben-Chorin, M., Diener, J., Averboukh, B., Polisski, G., and Koch, F. (1997) Symmetry of the electronic states of Si nanocrystals: an experimental study. *Phys. Rev. Lett.*, **79** (1), 119–122.
78. Fellah, S., Wehrspohn, R.B., Gabouze, N., Ozanam, F., and Chazalviel, J.-N. (1999) Photoluminescence quenching of porous silicon in organic solvents: evidence for dielectric effects. *J. Lumin.*, **80** (1–4), 109–113.
79. Mason, M.D., Sirbuly, D.J., Carson, P.J., and Buratto, S.K. (2001) Investigating individual chromopores within single porous silicon nanoparticles. *J. Chem. Phys.*, **114** (18), 8119–8123.
80. Mason, M.D., Credo, G.M., Weston, K.D., and Buratto, S.K. (1998) Luminescence of individual porous Si chromophores. *Phys. Rev. Lett.*, **80** (24), 5405–5408.
81. Chouket, A., Elhouichet, H., Oueslati, M., Koyama, H., Gelloz, B., and Koshida, N. (2007) Energy transfer in porous-silicon/laser-dye composite evidenced by polarization memory of photoluminescence. *Appl. Phys. Lett.*, **91** (21), 211902.
82. Ko, M.C., and Meyer, G.J. (1996) Dynamic quenching of porous silicon excited states. *Chem. Mater.*, **8** (11), 2686–2692.
83. Sweryda-Krawiec, B., Coffer, J.L., and Gancopadhahay, S. (2002) Time-resolved spectroscopic studies of the photoluminescence of nanoporous silicon: effects of Lewis acid/base exposure. *J. Cluster Sci.*, **13** (4), 637–645.
84. Harper, J., and Sailor, M.J. (1997) Photoluminescence quenching and the photochemical oxidation of porous silicon by molecular oxygen. *Langmuir*, **13** (17), 4652–4658.
85. Song, J.H., and Sailor, M.J. (1997) Quenching of photoluminescence from porous silicon by aromatic molecules. *J. Am. Chem. Soc.*, **119** (31), 7381–7385.
86. Palmer, R.K., Stein, D., Abrahams, E.S., and Anderson, P.W. (1984) Models of hierarchically constrained dynamics for glassy relaxation. *Phys. Rev. Lett.*, **53**, 958.

87 Medintz, I.L., Uyeda, H.T., Goldman, E.R., and Mattoussi, H. (2005) Quantum dot bioconjugates for imaging, labelling and sensing. *Nat. Mater.*, **4**, 435–446.

88 Dabbousi, B.O., Rodriguez-Viejo, J., Mikulec, F.V., Heine, J.R., Mattoussi, H., Ober, R., Jensen, K.F., and Bawendi, M.G. (1997) (CdSe)ZnS core-shell quantum dots: synthesis and characterization of a size series of highly luminescent nanocrystallites. *J. Phys. Chem. B*, **101** (46), 9463–9475.

89 Letant, S., and Vial, J.C. (1998) A luminescence versus temperature study of fresh and oxidized porous silicon layers under different atmospheres. *J. Appl. Phys.*, **84** (2), 1041–1046.

90 Mochizuki, Y., and Mizuta, M. (1995) Role of dangling bond centers on radiative recombination processes in porous silicon. *Appl. Phys. Lett.*, **67** (10), 1396–1398.

91 Mihalcescu, I., Ligeon, M., Muller, F., Romestain, R., and Vial, J.C. (1993) Surface passivation: a critical parameter for the visible luminescence of electrooxidised porous silicon. *J. Lumin.*, **57**, 111–115.

92 Ghulinyan, M., Gelloz, B., Ohta, T., Pavesi, L., Lockwood, D.J., and Koshida, N. (2008) Stabilized porous silicon optical superlattices with controlled surface passivation. *Appl. Phys. Lett.*, **93** (6), 061113.

93 Salhi, B., Gelloz, B., Koshida, N., Patriarche, G., and Boukherroub, R. (2007) Synthesis and photoluminescence properties of silicon nanowires treated by high-pressure water vapor annealing. *Phys. Status Solidi A-Appl. Mater.*, **204** (5), 1302–1306.

94 Petrova-Koch, V., Muschik, T., Kux, A., Meyer, B.K., Koch, F., and Lehmann, V. (1992) Rapid thermal oxidized porous Si- the superior photoluminescent Si. *Appl. Phys. Lett.*, **61** (8), 943–945.

95 Baldwin, R.K., Pettigrew, K.A., Ratai, E., Augustine, M.P., and Kauzlarich, S.M. (2002) Solution reduction synthesis of surface stabilized silicon nanoparticles. *Chem. Commun.*, (17), 1822–1823.

96 Buriak, J.M., and Allen, M.J. (1998) Photoluminescence of porous silicon surfaces stabilized through Lewis acid mediated hydrosilylation. *J. Lumin.*, **80** (1–4), 29–35.

97 Canaria, C.A., Lees, I.N., Wun, A.W., Miskelly, G.M., and Sailor, M.J. (2002) Characterization of the carbon-silicon stretch in methylated porous silicon – observation of an anomalous isotope shift in the FTIR spectrum. *Inorg. Chem. Commun.*, **5**, 560–564.

98 Gupta, P., Dillon, A.C., Bracker, A.S., and George, S.M. (1991) FTIR studies of H_2O and D_2O decomposition on porous silicon. *Surf. Sci.*, **245**, 360–372.

99 Ogata, Y., Niki, H., Sakka, T., and Iwasaki, M. (1995) Oxidation of porous silicon under water-vapor environment. *J. Electrochem. Soc.*, **142** (5), 1595–1601.

100 Ogata, Y., Niki, H., Sakka, T., and Iwasaki, M. (1995) Hydrogen in porous silicon – vibrational analysis of SiHx species. *J. Electrochem. Soc.*, **142** (1), 195–201.

101 Rivillon, S., Brewer, R.T., and Chabal, Y.J. (2005) Water reaction with chlorine-terminated silicon (111) and (100) surfaces. *Appl. Phys. Lett.*, **87** (17), 173118.

102 Bock, J., and Su, G.-J. (1970) Interpretation of the infrared spectra of fused silica. *J. Am. Ceram. Soc.*, **53** (2), 69–73.

103 Rao, S., Mantey, K., Therrien, J., Smith, A., and Nayfeh, M. (2007) Molecular behavior in the vibronic and excitonic properties of hydrogenated silicon nanoparticles. *Phys. Rev. B*, **76** (15), 155316.

104 Boukherroub, R., Wayner, D.D.M., Lockwood, D.J., and Canham, L.T. (2001) Passivated luminescent

porous silicon. *J. Electrochem. Soc.*, **148** (9), H91–H97.

105 Boukherroub, R., Morin, S., Wayner, D.D.M., and Lockwood, D.J. (2000) Thermal route for chemical modification and photoluminescence stabilization of porous silicon. *Phys. Status Solidi A-Appl. Res.*, **182** (1), 117–121.

106 Tsu, R., Shen, H., and Dutta, M. (1992) Correlation of Raman and photoluminecence spectra of porous silicon. *Appl. Phys. Lett.*, **60** (1), 112–114.

6
Chemistry of Porous Silicon

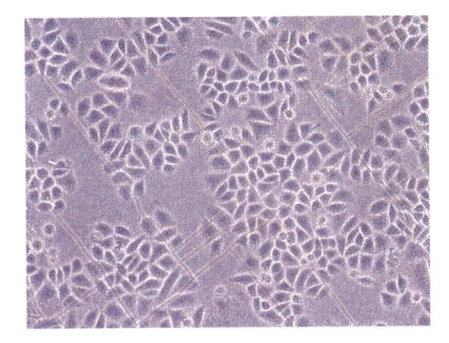

Figure 6.1
Human HeLa cells (a cervical cancer cell line) grown on a cracked porous photonic crystal. Surface chemistry is a critical factor determining biocompatibility, structural integrity, stability, and resorption rates of all biomaterials.

The reactivity of porous silicon is dominated by the chemistries of the silicon–hydrogen bond and the silicon–silicon bond. Both of these species are competent reducing agents, as was described in Chapter 1 in the context of the corrosion reactions at work in the formation of porous silicon. Both Si–H and Si–Si are able to reduce water, and, in the case of Si–Si bonds, the reduction reaction can generate new Si–H species in addition to oxides of silicon. Much of the chemistry of porous silicon has been developed to minimize or to harness these oxidation reactions – for example, one of the key limitations of optical biosensors based on porous silicon is the oxidation and subsequent dissolution of the mesoporous matrix in aqueous buffers, which leads to zero point drift and reduces the ultimate sensitivity of the devices [1]. By contrast, the use of porous silicon as an *in vivo* drug delivery or imaging material relies on the ability of the material to degrade into harmless, biocompatible constituents [2–4]. The most common approach

Porous Silicon in Practice: Preparation, Characterization and Applications, First Edition.
Michael J. Sailor.
© 2012 Wiley-VCH Verlag GmbH & Co. KGaA. Published 2012 by Wiley-VCH Verlag GmbH & Co. KGaA.

to stabilizing porous silicon is to purposely oxidize it. Usually performed at high temperatures in order to anneal and stabilize the Si–O bonds, oxidized material can be further modified using conventional silanol chemistries that are commonly employed to modify silica surfaces.

Several methods have been developed to chemically modify the freshly etched, hydrogen-terminated porous silicon surface that do not involve oxides of silicon. For example, the reductive power of porous silicon can be harnessed to spontaneously reduce many metal salts to their elemental forms, providing a route into interesting porous silicon/metal composites. Silicon-carbon bonds are much less polar than Si–O bonds, and reactions that form surface Si–C bonds show much greater resistance to attack by nucleophiles such as water or hydroxide. This chapter covers the important chemical transformations of porous silicon, with emphasis on modification of the material to improve its stability or to allow attachment of molecules.

6.1
Oxide-Forming Reactions of Porous Silicon

With its high surface area and reactive Si–H and Si–Si moieties, porous silicon is particularly susceptible to air or water oxidation. Once oxidized, nanophase SiO_2 readily dissolves in aqueous media [1], and the process is accelerated by surfactants or nucleophiles [5, 6]. A variety of chemical or electrochemical oxidants can be used to oxidize porous silicon, although the simplest oxidant is air. Air oxidation produces different types of surface species, depending on the temperature at which the reaction is performed and the humidity of the air [7]. The Si–Si bond is weaker than the Si–H bond (Table 1.1), and mild oxidation tends to attack the Si–Si bonds preferentially.

6.1.1
Temperature Dependence of Oxidation Using Gas-Phase Oxidants

The reactions given in Equations 6.1–6.3 are representative of the type of material produced in different temperature regimes during air or $O_{2(g)}$ oxidation. The rate of the reaction is highly dependent on the temperature. Room-temperature oxidation produces a fairly thin oxide layer within a few hours, which grows over the course of several months (probably assisted by traces of stronger oxidants such as ozone and nitrogen oxides in the atmosphere). Oxidation at 900 °C is sufficient to completely convert the porous silicon skeleton to silicon oxide, although the length of time needed to accomplish this transformation depends on the type of sample: microporous silicon will typically convert within 1 h, mesoporous silicon requires 3 h [8], and the conversion of macroporous silicon may not be complete even after 12 h. This is due to the thickness of the silicon features and the rate of diffusion through the oxide layer; for example, a macroporous silicon sample can have crystalline silicon domains several microns thick, which will oxidize quite slowly, even at 900 °C.

Atmospheric oxidation at temperatures below ~200 °C also generates surface hydroxy species, and adsorbed water is observed in the infrared spectrum of such samples that have been exposed to humid air. The presence of water vapor during oxidation is an important determinant of the quantity of surface hydroxy species in the product oxide [7].

$$H_2Si\text{-}SiH_2 + O_2 \xrightarrow{25°C} H\text{-}Si\text{-}O\text{-}Si\text{-}H \quad (6.1)$$

$$HSi\text{(Si)}_3 + O_2 \xrightarrow{60-100°C} (OH)Si(OSi)_3 \quad (6.2)$$

$$H_2Si\text{-}SiH_2 + O_2 \xrightarrow{200-900°C} \text{(silicon oxide network)} \quad (6.3)$$

6.1.2
Thermal (Air) Oxidation

Thermal oxidation is employed by the microelectronics industry to produce high quality oxides on silicon, and this also works well with porous silicon. The most common approach to oxidizing porous silicon is to heat it in a tube furnace (Lindberg Blue M, Thermo Scientific, www.thermo.com) or a box oven (Thermolyne FD1535M, Thermo Scientific, www.thermo.com), in air, for several minutes to several hours. The tube furnace can be fed with pure or humidified oxygen to better control the reproducibility of the process. Thermal stresses can cause sample cracking, so use care when inserting or removing samples into/from a hot furnace. A preferred approach is to insert the sample with the furnace at room temperature, ramp it up to the desired temperature (10 °C min^{-1}), hold at the set point for the prescribed period, and then let it cool to room temperature ($5\,°C\,min^{-1}$) prior to sample removal.

A rapid thermal oxidation (RTO) process will better preserve the fine-grained, crystalline silicon domains in porous silicon [9]. An RTO instrument uses high intensity visible light to heat the sample at very precise temperatures for very short periods of time (typically less than 1 min). The AccuThermo AW410 from Allwin21 corporation (www.allwin21.com) is representative. A typical protocol is to heat the porous silicon sample at a rate of $200\,°C\,s^{-1}$, hold at $900\,°C$ for 30 s, and then cool at a rate of $80\,°C\,s^{-1}$ [9]. This procedure generates a thinner oxide than the slower processes described above.

6.1.3
Ozone Oxidation

Ozone oxidation, usually performed at room temperature, forms a more hydrated oxide (Equation 6.4) than the thermal oxidation procedure. This preparation is quite hydrophilic, and it also readily dissolves in aqueous media [1]. Milder chemical oxidants, such as dimethyl sulfoxide (DMSO) [10] or pyridine [11] can also be used. Mild oxidants tend to produce thin oxide layers, particularly if the reaction is performed at a low temperature or for a short period of time.

$$\text{Si}_3\text{Si-H} + \text{O}_3 \xrightarrow{25\,°C} \text{Si}_3\text{Si-O-Si(OH)-O-Si} \quad (6.4)$$

The fragile nature of porous silicon is directly related to the strain that is induced in the film during electrochemical preparation [12], and the volume expansion that accompanies thermal oxidation can also introduce strain. Mild chemical oxidants tend to reduce strain in the films, because they attack porous silicon preferentially at Si–Si bonds that are the most strained, and hence most reactive [13]. Thus solution-borne, room temperature oxidants can relieve strain in the porous silicon layer and actually improve its mechanical stability somewhat. These factors are negated if the oxidant generates a gaseous by-product, because bubbles generated within the porous matrix can generate strong, destructive capillary forces. Stronger oxidants such as nitrate are very effective in generating surface oxide indiscriminantly, although the electron transfer kinetics of the oxidant plays a significant role in the uniformity of the oxide. For example, nitric acid has been the oxidant of choice for chemical stain etching [14] as described in Chapter 2 (Experiment 2.5). Stain etched material enjoyed a bad reputation for many years due to the heterogeneous nature of the nitrate oxidation reaction, which led to patchy surfaces. The discovery of chemical oxidants with more favorable kinetics for the stain etch process, such as vanadium oxides, has led to much greater uniformity and reproducibility [14] (see Experiment 2.5); in particular for porous silicon particles made from silicon powders.

6.1.4
High-Pressure Water Vapor Annealing

Heating a porous silicon sample in the presence of high pressure water vapor gives a very stable oxide [15, 16], Equation 6.5. When applied to the appropriate sample, this method can generate material with a strong blue phosphorescence (quantum yield up to 23%) [17, 18]. The emissive centers in this case are probably oxide-derived. The apparatus to perform the high-pressure water vapor annealing reaction is a sealed, heated pressure vessel. One example is the Parr instruments (www.parrinst.com) model 4565-T-SS-M-115-VS.12-2000-4848. This is a stainless steel reactor with a PTFE (Teflon) compression gasket and a capacity of 100 ml. It can be heated to 350 °C and can withstand pressures up to 2000 psi (13.7 MPa). To perform the reaction, the reactor is charged with the porous silicon sample and deionized water, sealed, and then heated to 260 °C for 3 h. The pressure inside the vessel will reach between 400–600 psi, depending on the temperature. The vessel is allowed to cool to room temperature before opening.

$$Si + 2H_2O \rightarrow SiO_2 + 2H_2 \tag{6.5}$$

6.1.5
Oxidation in Aqueous Solutions

As mentioned above, water is a competent oxidant for porous silicon, and the reaction with water vapor generates an oxide layer even at room temperature [19]. If the reaction is performed in liquid water, this oxide can dissolve. Beginning at pH values >7, hydroxide ions attack both Si–H and Si–O surface species; complete dissolution of a microporous layer occurs within a few minutes at pH 10. This reaction occurs by an associative mechanism at the silicon atom [20]; attack by a nucleophile generates a 5-coordinate intermediate which then induces Si–Si bond cleavage. In basic solutions the nucleophile is usually hydroxide (OH^-, Equation 6.6), but other nucleophiles such as amines can also perform the function [21]. The rate of nucleophilic attack increases if the silicon atom at the surface has electron-withdrawing substituents, like oxygen, and so oxidation of silicon is self-accelerating in aqueous media. The Si–O bonds are susceptible to hydrolysis, and dissolution occurs via Equation 6.7. The rate of oxide dissolution is slow in acidic solutions. This allows the formation of a stable oxide if porous silicon is oxidized in the presence of a mineral acid, as described in Section 6.1.6.

$$\text{Si-Si + OH}^- \longrightarrow \left[\text{Si-Si} \cdots \text{OH} \right] \xrightarrow{+ H_2O}_{- H_2} \text{Si-O-Si} + OH^- \tag{6.6}$$

$$SiO_2 + 2\ OH^- \rightarrow SiO_2(OH)_2^{2-} \tag{6.7}$$

6.1.5.1 Aqueous Oxidation Induced by Cationic Surfactants

Although dissolution of porous silicon or its oxides is slow in acidic solutions, various additives can increase the rate significantly. In particular, cationic surfactants cause rapid degradation and dissolution of porous silicon in aqueous media, even at acidic pH values (pH < 4) [6]. The origin of this unusual behavior lies in the interaction of the charged headgroup on the surfactant with the electronic structure of porous silicon in the presence of water.

The role of the cationic surfactant is thought to involve stabilization of negative charge at the porous silicon surface by the positively charged surfactant. Cationic surfactants polarize negative charge near the surface of the semiconductor, inducing more hydridic character in the Si–H surface species and enhancing the electrophilic nature of the silicon atoms at the surface (Figure 6.2) [6, 22]. The increased reactivity makes the surface more susceptible to attack by water.

6.1.6
Electrochemical Oxidation in Aqueous Mineral Acids

Electrochemical anodization of a porous silicon sample in a mineral acid, such as aqueous H_2SO_4, yields a fairly stable oxide, Equation 6.8 [23]. This procedure generates a hydrophilic oxide, enabling the incorporation and adsorption of hydrophilic drugs or biomolecules within the pores. If aqueous anodization is performed in the presence of Ca^{2+} ion, a calicified form of porous silicon results [24]. This formulation has been shown to be bioactive and is of particular interest for *in vivo* applications. Although it is enhanced by passing a DC electric current through the sample, oxidative calcification can also occur in the absence of an external current source [25, 26].

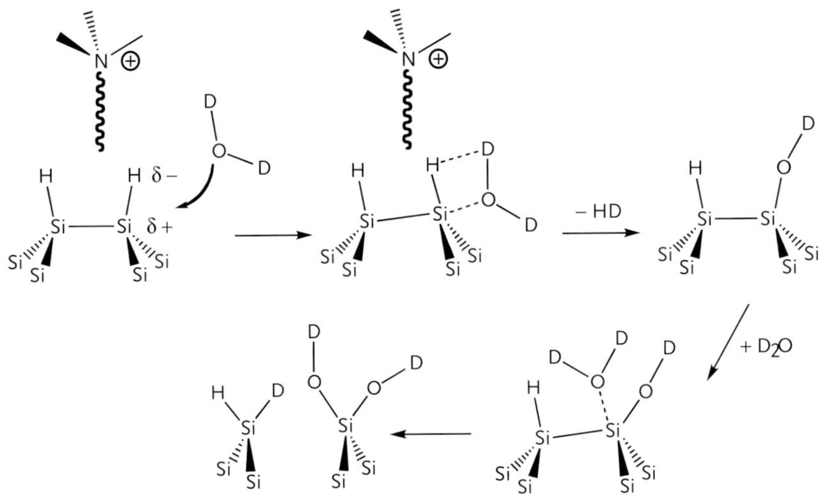

Figure 6.2
Mechanism of aqueous oxidation of porous silicon by water, catalyzed by cationic surfactants. Conclusions from a deuterium isotope tracer study are shown [6]. In this mechanism, the cationic surfactant catalyzes nucleophilic attack at a silicon atom by water. Elimination of hydrogen gas results in attachment of a hydroxy species to the surface. Attack by a second water molecule leads to cleavage of an adjacent Si–Si bond and transfer of hydrogen from the incoming water molecule to the silicon surface atom.

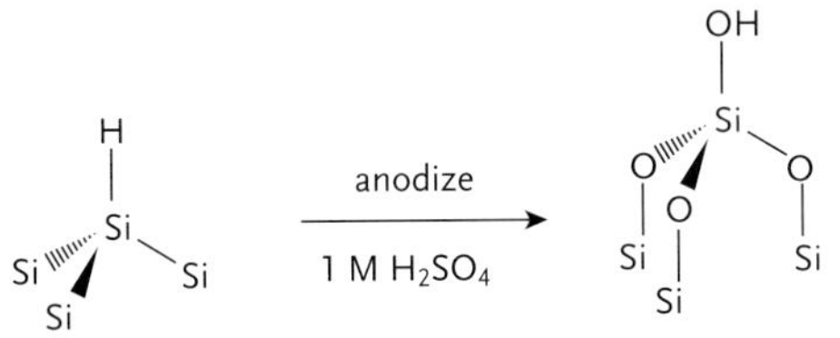

$$\text{(6.8)}$$

6.1.7
Oxidation by Organic Species: Ketones, Aldehydes, Quinones, and Dimethylsulfoxide

The reducing ability of surface Si–H species has also been observed with organic oxidants such as benzoquinone (Equation 6.9) [27] and various ketones and aldehydes [28], and it is similar to molecular silane chemistry in this regard [29]. The reduced organic fragment often shows up as a surface-bound species, and this reaction can be used to functionalize porous silicon [27].

[Structural equation (6.9) showing reaction of Si-H surface species with benzoquinone to give Si-O-aryl-OH surface species]

(6.9)

Dimethyl sulfoxide (DMSO) is also a mild organic oxidant for porous silicon [10]. When porous silicon is exposed to DMSO, an oxide layer is observed to grow, but the total concentration of H species on the porous silicon surface remains constant. Deuterium tracer studies indicate that the primary oxidation route with DMSO is cleavage of Si–Si, rather than Si–H bonds (Equation 6.10). Unlike the RTO process, this mild oxidation reaction generates a defective porous silicon surface that displays significantly weaker photoluminescence than the starting material. These defects (surface dangling bond states) can be suppressed and photoluminescence preserved if the reaction is carried out in the presence of a hydrogen atom donor (butylated hydroxytoluene, BHT), which hydrogenates and passivates the dangling bond states [10]. The byproduct of the reaction is dimethylsulfide, which undergoes no further reaction with porous silicon (Equation 6.10).

[Structural equation (6.10) showing Si-Si surface bond + DMSO (H_3C-S(=O)-CH_3) → Si-O-Si surface bridge + $(CH_3)_2S$]

(6.10)

6.1.8
Effect of Chemical Oxidation on Pore Morphology

When a porous silicon skeleton oxidizes, each silicon atom picks up two oxygen atoms. The volume expansion associated with this chemical reaction effectively reduces the free volume of the pores and decreases their average diameter, as we saw in the BET porosimetry measurements of Figure 5.19. The oxidation-induced pore shrinkage phenomenon can be harnessed to physically trap molecules and nanoparticles in porous silicon. For example, iron oxide [30–33] and cadmium selenide [3] nanoparticles have been oxidatively trapped in porous silicon. As was mentioned above,

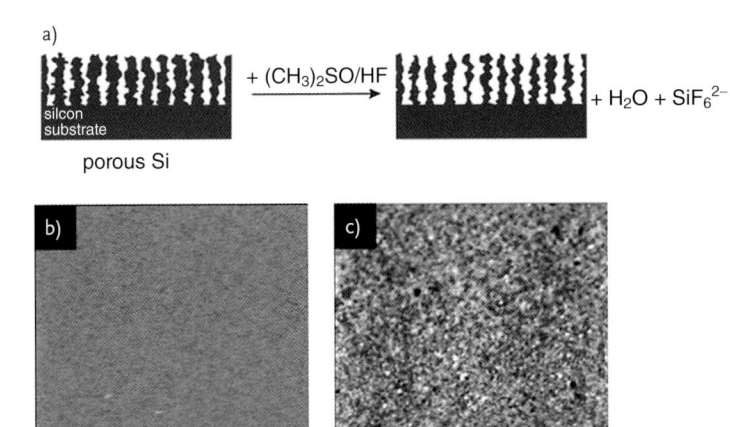

Figure 6.3
(a) Pore expansion chemistry using dimethyl sulfoxide/aqueous HF mixtures. Dimethyl sulfoxide acts as a mild oxidant that slowly converts silicon to silicon dioxide. As it forms, the oxide is removed by HF$_{(aq)}$. The overall effect is to expand the pores and increase the porosity of the porous silicon matrix. (b) and (c) Atomic force micrograph images (tapping mode) of porous silicon (p-type) surfaces before (a) and after (b) pore expansion by immersion in a 9:1 (v/v) 49% aqueous HF:DMSO solution for 30 min. The images show an increase in the mean pore size upon chemical etching. The porosity of the sample has increased from 50% to 60%.

oxide generation also induces strain in the matrix, which can be manifested in cracking of the film to generate new pores or fissures.

If a mechanism exists to remove the oxide as it is formed, the pores can be expanded. For example, treatment of porous silicon with dimethyl sulfoxide solutions containing HF will generate larger pores by simultaneously oxidizing the porous silicon skeleton and dissolving the oxide that is generated (Equation 6.11 and Figure 6.3) [13, 34]. For this procedure, a freshly-etched film is soaked in a solution consisting of aqueous HF and DMSO. A typical recipe calls for soaking in a 4:1:1(v/v) DMSO:48% aqueous HF:ethanol mixture [34]. The process can be conveniently monitored by optical reflectivity spectroscopy, as described in Chapter 5, although there is no *in situ* means of assessing the pore size during the process.

Aqueous solutions of bases such as KOH can also be used to enlarge the pores after etching [35]. Basic aqueous solutions such as KOH tend to dissolve the smaller, microporous features of a porous silicon film first, although strongly basic solutions will dissolve a micro- or meso-porous sample completely (Experiment 5.1).

$$\text{(scheme: Si-H surface species} \xrightarrow{\text{DMSO}} \text{Si-O-Si bridged species} \xrightarrow{\text{HF}} \text{SiH}_2 \text{ species)} \tag{6.11}$$

Since the pore dimensions in porous silicon depend on the electrochemical current used in preparing the material (Chapter 2), larger pores can be obtained simply by increasing the current density of an etch [36]. Much work has focused on this approach, and with the right combination of current density, HF concentration in the etch bath, and wafer resistivity, material with pores sufficient to admit large biomolecules such as bovine serum albumin or immunoglobulin G have been obtained [1, 37, 38]. The electrochemical approach to generating large pores is convenient, although it is also somewhat problematic since the high current densities needed to obtain mesopores (e.g., on highly doped p-type silicon substrates) are close to the electropolishing regime – leading to cracking and flaking of the films. Various chemical "additives" and non-aqueous solvents have been used in the electrochemical etch of porous silicon to access mesoporous morphologies without inducing excessive mechanical instability [39–42]. The advantage of using a mild chemical oxidant coupled with, or after the electrochemical etch is that it can reduce mechanical strain in the film because the chemical oxidant attacks porous silicon preferentially at regions of the film where the most strain exists. Such a mechanism is not expected to hold for electrochemical dissolution, where pore formation is determined more by fluoride ion concentration and local changes in silicon conductivity than by strain [43].

In addition to its utility in opening the pores of a pre-formed porous silicon layer, DMSO/HF solutions can be used to remove residual porous silicon from a composite. For example, the silicon portion of a carbon-infiltrated mesoporous silicon film can be extracted by soaking the film in a 4:1:1(v/v) DMSO:48% aqueous HF:ethanol mixture [34]. The process can take several hours to reach completion for a mesoporous sample.

6.2
Biological Implications of the Aqueous Chemistry of Porous Silicon

Silicon is an essential trace element that is linked to the health of bone and connective tissues [44]. The chemical species of relevance to the toxicity of porous silicon are silane (SiH_4) and dissolved oxides of silicon. As discussed in Chapter 5, the surface of porous silicon contains various quantities of Si–H, SiH_2, and SiH_3 – species that can convert to silane under the appropriate conditions [45, 46]. Silane is a highly reactive gas that spontaneously combusts in air via Equation 6.12, making it a significant explosion hazard. It is considered toxic; inhalation can irritate the respiratory tract and cause headache and nausea. The hydrolysis of silane in the body tissues forms silicic acid and hydrated silica, and inhalation toxicity has been demonstrated in mice at 1000 ppm [47], with death occurring at concentrations of 10 000 ppm [48]. There is a story from the early days of porous silicon research that a sample of porous silicon set off the silane alarms when it was brought into a semiconductor fabrication facility. The quantity of silane that can potentially be generated from the porous silicon samples prepared at the scale described in this book is not large. For example, the amount of silicon contained in a 20-μm thick, 50% porous silicon film prepared using the Standard etch cell (1.2 cm^2) is 2.8 mg; if the entire porous film were converted to silane, it would produce only 3.2 mg of silane. If released into a confined space, this small quantity could pose an inhalation hazard (if it were released into 1 m^3 of air, the concentration would be 2.7 ppm). Porous silicon in direct contact with mammalian cells can generate cytotoxicity associated with localized release of silane [46].

$$SiH_4 + 2\, H_2O \rightarrow SiO_2 + 4\, H_2 \tag{6.12}$$

Like silane, the native SiH_x species on the porous silicon surface readily oxidize in aqueous media. Silicon itself is thermodynamically unstable toward oxidation, and even water has sufficient oxidizing potential to make this reaction spontaneous (Equation 6.5). The passivating action of SiO_2 and Si–H (for samples immersed in HF solutions) makes the spontaneous oxidation of silicon in water kinetically slow. Once oxidized, however, the resulting porous silicon oxide can readily dissolve in water to expose fresh silicon. Because of its highly porous nanostructure, oxidized porous silicon can release relatively large amounts of silicon-containing species into solution in a short time. The soluble forms of SiO_2 exist as various silicic acid compounds, with the orthosilicate (SiO_4^{4-}) ion as the basic building block (Equation 6.13; see Chapter 1 for a more extensive discussion of the aqueous chemistry). These oxides can be toxic in high doses, generally due to precipitation of the silicate ion in the kidneys, which leads to renal failure [49–51].

$$SiO_2 + 2\, H_2O \rightarrow Si(OH)_4 \tag{6.13}$$

Because the body can handle and eliminate silicic acid, the important issue with porous silicon-based drug delivery systems is the rate at which they degrade and resorb [24–26, 52, 53]. The work of Bayliss, Canham, and others established the relatively low toxicity of porous silicon in various mammalian cells and live animals [54–60]. The low toxicity and solubility of the degradation byproducts of porous silicon have generated much interest for its use in controlled drug delivery systems.

Surface chemistry plays a large role in controlling the rate of degradation of porous silicon *in vivo*. Immediately after silicon is electrochemically etched, the surface is covered with reactive hydride species. This chemical functionality provides a versatile starting point for various reactions that can control the dissolution rate in aqueous media, allow attachment of targeting species, and control release of drugs. Presently, the two most important modification reactions are chemical oxidation (which we discussed above) and grafting of Si–C species.

6.3
Formation of Silicon–Carbon Bonds

6.3.1
Thermal Hydrosilylation to Produce Si–C Bonds

Carbon directly bonded to silicon yields a very stable surface species. First recognized by Chidsey [61] in the early 1990s, Si–C bonded species possess greater kinetic stability relative to Si–O due to the low electronegativity of carbon. Silicon can readily form 5- and 6-coordinate intermediates, and an electronegative element such as oxygen enhances the tendency of silicon to be attacked by nucleophiles. The most ubiquitous reaction used to form a Si–C bond to hydrogen-terminated porous silicon is hydrosilylation (Equation 6.14), a reaction first demonstrated by Buriak [62–64] and elaborated by Boukherroub, Chazalviel, Lockwood, and many others [53, 65–71]. Hydrosilylation involves addition of Si–H across a C–C double or triple bond. The C–C multiple bond of the alkene (C–C double bond) or alkyne (C–C triple bond) is usually at one end of the hydrocarbon chain, a so-called terminal bond. On porous silicon, the reaction is promoted by heat [68], light [72, 73], mechanical scribing [74], or by Lewis acid catalysis [63, 64].

(6.14)

Thermal hydrosilylation provides a means to place a wide variety of organic functional groups on a porous silicon surface, including carboxylic acid or ester groups that then allow further chemical modification [63]. The main requirement of the reaction is that the surface contains Si–H species so they can react with the alkene or alkyne. Thus it is important to use freshly etched porous silicon and to exclude oxygen and water from the reaction mixture. This is most easily done on a Schlenk line.

6.3.2
Working with Air- and Water-Sensitive Compounds – Schlenk Line Manipulations

As with many of the reactions described in this chapter, thermal hydrosilylation requires inert-atmosphere manipulations. The most convenient way to perform these reactions is on a vacuum apparatus known as a Schlenk line. A great reference for all needed Schlenk and syringe techniques is *The Manipulation of Air-Sensitive Compounds*, by Shriver and Drezdzon [75]. The apparatus is available from chemistry glassware suppliers such as Chemglas (www.chemglas.com). The safe set-up, use and operation of a Schlenk line is beyond the scope of this book, and the interested reader should consult the Shriver book listed above for further details. In Experiment 6.1 we will give a representative Schlenk line procedure, using the thermal hydrosilylation reaction as an example.

EXPERIMENT 6.1: Thermal hydrosilylation of a porous silicon sample with 1-dodecene

In this experiment you will prepare a hydrophobic porous silicon sample by grafting a 12 carbon-long alkane chain to the freshly etched surface. The alkane imparts good stability and a high degree of hydrophobicity to the surface. You will also confirm that the organic group is attached to the porous silicon surface by a Si–C bond by treating the modified sample with HF. The freshly etched porous silicon sample can be prepared in Experiment 2.2.

Equipment/supplies:

Porous silicon sample (freshly etched)	This method will work with most porous silicon samples as long as they are freshly etched.
Schlenk line	Chemglas Airfree® product line: Manifold Vacuum, Inert Gas AF-0060 (www.chemglas.com). Note you will need additional adapters, liquid nitrogen traps, a source of inert gas (nitrogen or argon), liquid nitrogen, and flasks – see reference [75] for details. The vendor sells a complete vacuum system (AF-0300-1). The single-bank manifold (AF-0300-03) on this unit must be replaced with the dual-bank manifold (AF-00600) for Schlenk line manipulations.

Reaction vessel	Chemglas Airfree® product catalog number AF-0556-01. Drying Chamber, #40 O-Ring Joint, Lower Vessel 40 mm ID × 100 mm height, Airfree, #65 Clamp Size, #226 O-Ring Size (www.chemglas.com)
1-dodecene	Sigma Aldrich chemicals, catalog number 44148-10ML (www.sigmaaldrich.com)

Procedure:

1) Obtain a freshly etched porous silicon chip (Experiment 2.2 provides an example for a sample derived from p^{++}-silicon).

2) Place the sample face up in the reaction vessel. Seal the vessel and evacuate it for 5 min to ensure the sample has no residual ethanol.

3) Backfill the vessel with nitrogen and then add ~0.5 ml of 1-dodecene via syringe directly to the porous silicon surface.

4) Evacuate the vessel briefly to remove all nitrogen (do not evacuate too long or the 1-dodecene will evaporate), and close the valve on the vessel.

5) Heat the sealed vessel in a silicone oil bath to a temperature of 120 °C for 2 h.

6) Cool the vessel to room temperature, purge it with nitrogen and remove the sample.

7) Rinse the sample with ethanol and dry it thoroughly.

8) Obtain an infrared spectrum of the sample.

9) Rinse the sample in 1:1 aqueous 48% HF:ethanol, then rinse the sample with pure ethanol and dry it thoroughly.

10) Obtain a second infrared spectrum.

Results:

The infrared spectrum of the sample should show strong bands associated with the C–H stretching vibrations of the grafted organic species, and minimal oxide (Figure 6.4). The C=C double bond of the alkene goes away in the reaction. The purpose of steps 9 and 10 is to confirm that the organic group is covalently attached to the porous silicon sample. Very commonly, the small amount of oxidation that accompanies a Si–C grafting experiment will trap some organic starting material within the porous matrix. Since Si–O bonds are removed by HF, treatment of a modified porous silicon sample with ethanolic HF serves as a "litmus test" for Si–C grafted species. Silicon–carbon bonds are not attacked by HF.

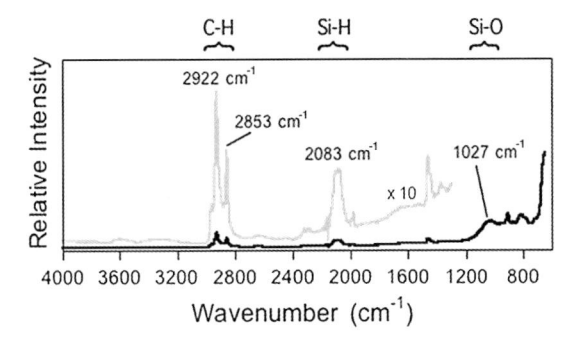

Figure 6.4
Fourier transform infrared (FTIR) spectrum of a porous silicon sample modified with a dodecane functionality by thermal hydrosilylation with 1-dodecene. Strong C–H bands associated with the organic group should be observed even after rinsing the sample with an ethanolic HF solution.

6.3.3
Classification of Surface Chemistry by Contact Angle

The surface of the dodecyl-terminated porous silicon sample prepared in Experiment 6.1 is quite hydrophobic. Figure 6.5 shows photographs of water drops on two porous silicon samples, comparing the hydrophobicity of the dodecyl-terminated surface with the thermally oxidized porous silicon surface. Both samples were prepared with the same etch parameters such that they had similar average pore size, porosity, and thickness. The water drop spreads out on the hydrophilic SiO_2 layer, whereas water is effectively repelled from the surface of the dodecyl-modified material, causing it to "bead up." The photograph illustrates one of the more common and readily applied measurements to characterize surface chemistry—water contact angles.

Contact angle is defined as the angle between a line tangent to the surface of a liquid drop and a solid, flat surface that the drop contacts. The tangent is drawn as close to the point of contact as possible. Table 6.1 provides representative contact angles for some of the porous silicon chemistries discussed in this chapter. It should be stressed that feature size plays a substantial role in the values of contact angles measured on porous surfaces [76–80], and the numbers in the table are meant to be illustrative for comparison of sample chemistries. For example, the contact angle on the dodecyl-modified sample shown in Figure 6.5 appears to be of the order of 90° rather than the 120° value listed in Table 6.1. This is due to a lower degree of functionalization that was used to prepare the sample in Figure 6.5. All samples measured in Table 6.1 were prepared using the same etch parameters and measured immediately after modification.

It is important to keep in mind that the thermal hydrosilylation reaction does not completely remove all Si–H species from the surface (Figure 6.4).

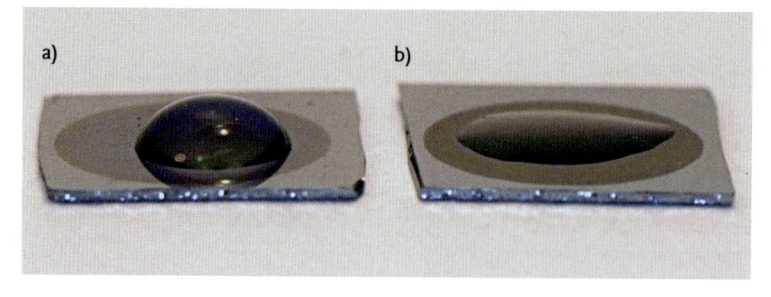

Figure 6.5
A pair of porous silicon samples, each with a drop of water placed on its surface. The sample (a) contains a hydrophobic surface chemistry (dodecane) that excludes water, while the sample (b) contains a hydrophilic surface (SiO_2). Water spreads out on the surface and wicks into the hydrophilic chemistry much more effectively.

The thermal reaction typically only reacts with a few percent of Si–H species on porous silicon [67], and even the more effective Lewis acid catalyzed reaction is less than 30% efficient [64]. This can be attributed to steric constraints; most organic molecules cannot fit into the very smallest voids in a porous silicon sample, and they are too large to attach to adjacent Si–H moieties on the porous silicon surface. In thermal hydrosilylation, the extent of surface coverage is often a race between the rate of hydrosilylation and the rate of oxidation of the porous silicon sample by adventitious water or oxygen.

6.3.4
Microwave-Assisted Hydrosilylation to Produce Si–C Bonds

One of the most convenient and efficient hydrosilylation routes is microwave-assisted hydrosilylation, which involves irradiation of a porous silicon sample that is immersed in neat alkene with a microwave source. This reaction is much more rapid than the conventional thermal reaction, and so it tends to minimize sample oxidation. Reaction efficiencies as high as 38% have been reported [67]. Due to the rapid conversion of microwave energy to heat in a conductive silicon wafer, and the high flammability of most alkenes, this reaction is quite hazardous. In the first demonstration of microwave-assisted hydrosilylation by Boukherroub, Ozanam, and Chazalviel, the porous silicon sample was attached to the silicon substrate and the microwave power and temperature of the reaction were carefully monitored [67]. A microwave source that was specifically designed for chemical reactions was used (Synthewave 402 monomode reactor, Prolabo, inc.).

Table 6.1 Contact angles of porous silicon surface chemistries[a].

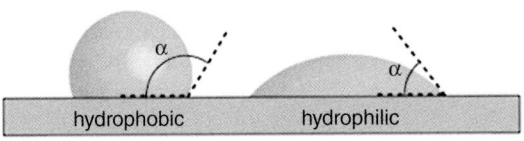

Process	Chemical reaction	Surface species	Contact angle α (degrees)
Native surface	–	Si–H	100
Ozone oxidation	Si–H + O$_3$	Si–O–Si, Si–OH	11
Thermal oxidation (600 °C, 90 min)	Si–H + O$_2$	Si–O–Si	20
Hydrosilylation with 1-dodecene	Si–H + CH$_2$=CH$_2$(CH$_2$)$_9$CH$_3$	Si–(CH$_2$)$_{11}$CH$_3$	120
Electrochemical methylation	Si–H + CH$_3$I	Si–CH$_3$	100
Ozone oxidation + dichlorodimethylsilane	(a) Si–H + O$_3$ (b) 2 Si–OH + Cl$_2$Si(CH$_3$)$_2$	Si–O–Si(CH$_3$)$_2$–O–Si	100
Thermal acetylation (300 °C, 30 min)	Si–H + H–C≡C–H =>	"Si–C"	53
Thermal acetylation (485 °C, 30 min)	Si–H + H–C≡C–H	"Si–C"	80
Thermal acetylation (500 °C, 30 min)	Si–H + H–C≡C–H	"Si–C"	76

a) Samples were from p^{++}-wafers, etched at 40 mA cm^{-2}. Porosity of all samples 75%. Surface species shown in the table represent idealized surface chemistry. Many of the reactions also generate oxide concomitant with the modification reaction. The presence of a surface oxide tends to reduce the measured value of contact angle.

The author's laboratory has used a commercial microwave oven designed for the consumer market. In one case, explosion of a porous silicon chip that had been immersed in neat dodecene resulted in instantaneous and forceful opening of the oven door. Microwaves couple much less efficiently into porous silicon than they do into a bulk silicon wafer, and the reaction is less hazardous with lift-offs or with micron-size porous silicon powders. However, the use of a microwave reactor designed for chemical reactions, and that has adequate explosion safety protection is recommended for either chip or powder-based reactions. The reaction should be performed in an inert atmosphere, the alkene should be degassed prior to reaction, and the amount of alkene and porous silicon should be sufficiently small. To illustrate the hazard, if 200 g of dodecene suddenly vaporizes at 300 °C in a 30-l container (the volume of a small microwave oven), the pressure within the vessel will triple. Under the same conditions, vaporization of 2 g of dodecene will only increase the pressure by ~2%. Of course, if the organic vapor in either of these two examples mixes with air and ignites, the resulting explosion will generate significant heat, pressure, and possible damage.

6.3.5
Chemical or Electrochemical Grafting to Produce Si–C Bonds

As an alternative to hydrosilylation, covalently attached layers can be formed on porous silicon surfaces using Grignard and alkyl- or aryl-lithium reagents [61, 81–87]. Electrochemical oxidation of methyl-Grignards on porous silicon yields dense monolayers of methyl species [88]. The electrochemical reduction of organohalides (Equation 6.15) has also been demonstrated as an effective grafting technique [89–92]. These reactions are performed in non-aqueous solvents, and care must be taken to avoid the introduction of water to the reaction. The Standard etch cell has a groove on the top that allows the mating of a glass cap via an O-ring fixture, shown

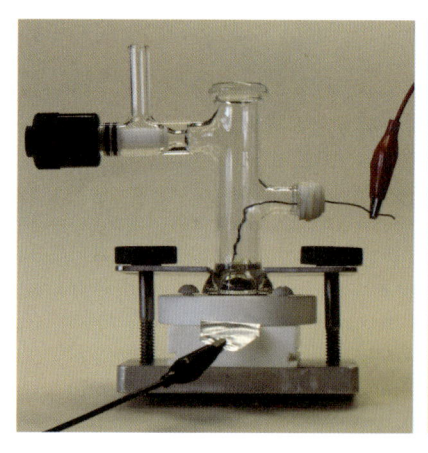

Figure 6.6
Photograph of a cell used for anaerobic electrochemistry on porous silicon samples. A glass airless cap, fitted with an electrical feed--through and a Teflon valve is mounted on the top of a Standard etch cell via an O-ring seal. This allows the addition of electrolytes or reactants to perform air- or water-sensitive reactions on a porous silicon sample. The wire feed-through is attached to a platinum loop that is immersed in the electrolyte and serves as a counter electrode. The clamp holding everything together is made of stainless steel (top piece) and aluminum (bottom piece). The top of the glass cap has an optical flat that allows acquisition of optical reflectivity spectra from the porous silicon sample *in situ*.

pictorially in Figure 6.6. To perform electrochemical modification, a platinum counter-electrode is fed through a sealed hole in the glass cap. The electrochemical approach allows attachment of a methyl group to the silicon surface, which it is not possible to achieve by hydrosilylation. Because the products show significant stability in aqueous media, hydrosilylation and electrochemical grafting are useful reactions for the preparation of biointerfaces.

$$\begin{array}{c} \text{H} \\ | \\ \text{Si}\overset{\text{Si}}{\diagdown}\text{Si} \\ | \\ \text{Si} \end{array} \xrightarrow[\text{RCH}_2\text{X}]{+e^-} \begin{array}{c} \text{CH}_2\text{R} \\ | \\ \text{Si}\overset{\text{Si}}{\diagdown}\text{Si} \\ | \\ \text{Si} \end{array} \tag{6.15}$$

R = H, $(CH_2)_n CH_3$, $(CH_2)_3 CO_2 R$

X = I, Br

As with the hydrosilylation reaction, electrochemical grafting reactions do not provide 100% surface coverage; they merely decorate the surface with the functional group. Thus infrared spectra show a large amount of surface Si–H groups remaining after the reactions. The electrochemical method allows one to minimize residual Si–H species by "endcapping" the surface with small methyl groups following modification with a functional species (Equation 6.16) [92]. The endcapping reaction can also be performed on a hydrosilylated porous silicon surface. Methyl endcapping of porous silicon surfaces significantly improves their stability in aqueous media [92].

$$\tag{6.16}$$

The reason these modification reactions impart such stability to porous silicon can be ascribed to two factors: (i) the low polarity of the Si–C bond makes it kinetically inert toward attack by nucleophiles such as water or amines, and (ii) the attached organic species (typically a hydrocarbon chain 8 or more CH_2 units long) is sufficiently hydrophobic that water is excluded from the immediate vicinity of the attachment point. For example, porous silicon modified by hydrosilylation of dodecene (generating a 12-carbon aliphatic hydrocarbon on the surface) is stable to hydroxide solutions of pH >10, whereas unmodified (H-terminated) porous silicon dissolves rapidly under such conditions [64]. If the grafted species possesses a shorter chain or a hydrophilic group such as an ester, the modified porous silicon surface is noticeably less stable in water [92]. In addition, porous silicon modified by Si–C chemistry is still susceptible to air oxidation due to the small size and lack of polarity of molecular oxygen.

6.4
Thermal Carbonization Reactions

The search for functional porous silicon surfaces that are stable in aqueous media has been driven by interests in biosensor and drug delivery applications where longer-term stability (hours to months) is desired [2, 4]. The grafting reactions discussed in the previous section all suffer from the fact that reactive Si–Si and Si–H species are still present in the porous silicon material, and they are still accessible to aqueous solution. The Si–O bonds in oxidized porous silicon are likewise unstable in aqueous media due to hydrolytic attack. By contrast, silicon carbide (and to a lesser extent, silicon nitride) is a very stable material that could potentially withstand years of exposure to aqueous solutions without degradation. It was this idea that led Salonen and coworkers to develop a carbonization reaction akin to those that produce silicon carbides.

6.4.1
Thermal Degradation of Acetylene to form "Hydrocarbonized" Porous Silicon

Reaction of porous silicon with gas phase acetylene generates highly carbonized porous silicon that is very stable in aqueous media (Equation 6.17) [93–95]. Thermally carbonized porous silicon (TCPSi) has been extensively investigated by Salonen and coworkers, with many publications of relevance to drug loading and delivery [4, 96–101]. Although less well-defined than material prepared by the hydrosilylation route, "hydrocarbonized" porous silicon (TCPSi) is environmentally stable [102] and its chemical sensing capabilities have been demonstrated [103–107].

$$\text{Si-SiH}_2 + H-C\equiv C-H \xrightarrow{485°C} \text{Si-C-C-Si}$$

(6.17)

Salonen's thermal carbonization reaction is carried out by placing a freshly etched porous silicon sample in a tube furnace (Lindberg/Blue M), purging the tube with a constant flow of acetylene and nitrogen gas, and then increasing the temperature. Note that acetylene gas is highly flammable, and so it is safer to dilute the gas with nitrogen, or at a minimum to purge the tube with nitrogen before and after treatment if pure acetylene is to be used. In a typical preparation, nitrogen is flowed through the tube for

10 min at a flow rate of 1 l min^{-1}, acetylene is then introduced at a flow rate of 1 l min^{-1}, such that the total flow rate is 2 l min^{-1} of a 1:1 (by volume) mixture of acetylene and nitrogen. The temperature is then ramped to 450 °C (+50 °C min^{-1}) and maintained for 30 min. The gas is then switched back to pure nitrogen at a flow rate of 1 l min^{-1} and the tube furnace is allowed to cool to room temperature under the nitrogen flow. The resulting "hydrocarbonized" film generally remains porous. It is chemically composed of an ill-defined mixture of C–H, aromatic C=C, and C–OH species. The temperature of the pyrolysis reaction plays a key role in determining the degree of hydrophobicity or hydrophilicity of the resulting "hydrocarbonized" porous film [93, 108]. The reaction of acetylene with porous silicon at temperatures >500 °C can lead to significant carbonization [102], and higher temperatures or longer reaction times generate black films with a sooty appearance [103].

6.4.2
Thermal Degradation of Polymers to Form "Carbonized" Porous Silicon

One of the most common methods used to generate high surface-area carbons is pyrolysis of a carbon-containing resin or polymer, and this reaction will also work inside the pores of a porous silicon tempate [34]. Porous composites can be synthesized by thermal decomposition of poly(furfuryl alcohol), one of the more common polymers used as a precursor for amorphous carbons (Equation 6.18). The polymerization of furfuryl alcohol is acid-catalyzed, and the surface protons within an ozone-oxidized porous silicon sample (Equation 6.4) are sufficiently acidic to catalyze the reaction. The pyrolysis procedure (700 °C under flowing nitrogen for 5 h) yields carbonized bundles of fibers embedded in the porous silicon matrix. These materials still retain some porosity, although the average pore diameter decreases. The additional microporosity associated with the carbon matrix provides an increase in the total surface area. The presence of the microporous carbon matrix yields a tenfold increase in sensitivity for detection of organic vapors in optical films of the material [34]. Removal of the porous silicon template in a dimethyl sulfoxide solution of hydrofluoric acid (Equation 6.11) yields arrays of freestanding carbon nanofibers that replicate the porosity of the template (Figure 6.7).

Figure 6.7
Array of carbon nanofibers synthesized by thermal decomposition of poly(furfuryl alcohol) in a porous silicon template. The porous silicon template has been removed using the DMSO/HF processing step of Equation 6.11. The freestanding carbon fibers were prepared using the polymerization/pyrolysis procedure of Equation 6.18. This particular image is from a porous silicon double-layer template consisting of a high porosity layer atop a lower porosity layer. Image adapted from reference ([34]).

(6.18)

6.5
Conjugation of Biomolecules to Modified Porous Silicon

Biomolecules are attached to porous silicon surfaces for a variety of reasons: to impart selectivity to biosensors, to add homing molecules to target porous silicon nanoparticles to diseased tissues, to minimize biocompatibility issues associated with placing a foreign object in the body, to deliver drugs, and to reduce environmental fouling, to name just a few. The book *Bioconjugate Techniques* by Hermanson [109] is a leading reference on the chemical reactions that can be employed to link one molecule to another or to a surface. In the context of this chapter, the main strategy is to use a linker that has two different chemistries on either of its two ends. One chemistry sticks the linker to the porous silicon surface and the other sticks the linker to the molecule of interest. In general, an activating step is used that temporarily makes the surface linker more reactive, to allow formation of a covalent bond with the target molecule. The activating step makes use of a coupling agent; a classic example is 1-ethyl-3-[3-dimethylaminopropyl] carbodiimide hydrochloride (EDC).

6.5.1
Carbodiimide Coupling Reagents

EDC couples carboxyl groups to primary amines to form an amide bond. Thus the surface must have a carboxyl ($-CO_2H$) species and the target molecule must have a primary amine ($-NH_2$), or vice versa. An example is shown in Equation 6.19. The example given uses a porous silicon surface that has been modified by hydrosilylation of undecylenic acid (Equation 6.14), although any reaction that provides a free carboxy species should work. EDC is added to the surface first, because it reacts with the carboxylic acid on the porous silicon surface and generates a reactive intermediate (*O*-acylisourea). This intermediate is not particularly stable, and it reacts with the amine-containing species (in the example of Equation 6.19, the anti-cancer drug doxorubicin [22, 110]) that is added in a subsequent step. If the amine species is not added immediately, traces of water in the reaction will hydrolyze the EDC-activated surface to regenerate the carboxyl group. In order to temper the reaction and reduce its susceptibility to water, a slightly less reactive intermediate can be generated by addition of *N*-hydroxysulfosuccinimide (sulfo-NHS) to the EDC-activated surface instead of the target molecule [109]. This forms an ester that can still react with primary amines on our target molecule to form the desired peptide-linked product. Another example of a more stable intermediate is pentafluorophenol; an example is given in Equation 6.20. This particlular reaction was employed with hydrocarbonized porous silicon [111].

6 Chemistry of Porous Silicon

$$(6.19)$$

$$(6.20)$$

Due to solubility limitations, the EDC reaction must be performed in a polar solvent such as water. While an aqueous reaction medium is desirable if one wishes to attach a protein to the surface (proteins do not like solvents other than water), there are some experiments where water must be avoided. The EDC reaction can also be carried out in the polar, aprotic solvent dimethylformamide (DMF), as shown in Equation 6.20. A related carbodiimide reagent, N,N'-dicyclohexylcarbodiimide (DCC) is a neutral molecule, so it readily dissolves in less polar organic solvents such as dichloromethane. A generic reaction with this reagent is given in Figure 6.8.

6.5.2
Attachment of PEG to Improve Biocompatibility

There are two main considerations when linking molecules to porous silicon surfaces for biological applications. Stability in aquous media is one of them, which we have already discussed. The other is non-specific binding of unwanted proteins and other interfering species to the surface. Non-specific binding leads to fouling and degradation of sensitivity in biosensors, and in drug delivery materials it plays a key role in determining the biocompatibility of the fixture or device. Common coatings that minimize non-specific binding are carbohydrates (like dextran) and polyethylene

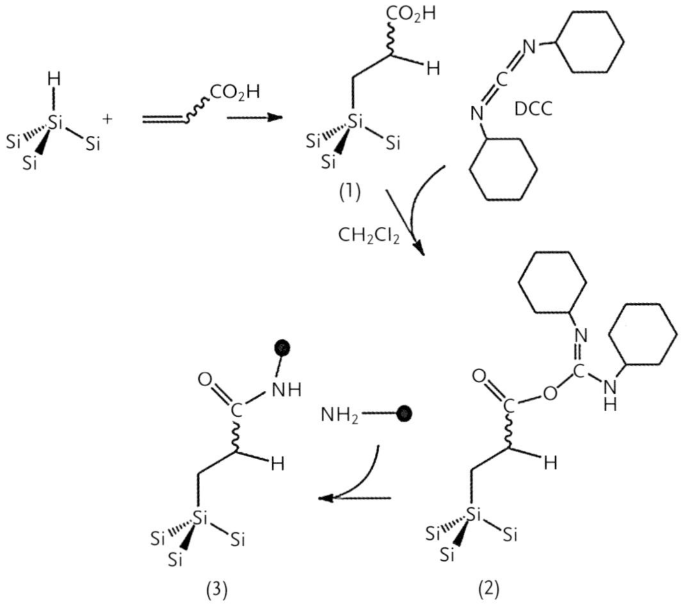

Figure 6.8
Non-aqueous route to attach a biomolecule to porous silicon in three steps. (1) The first step modifies the fresh porous silicon surface via hydrosilylation, grafting the –CO$_2$H functional group to the surface via a Si–C bond. (2) The surface is then reacted with the linking agent N,N'-dicyclohexylcarbodiimide (DCC) dissolved in the nonreactive solvent dichloromethane, CH$_2$Cl$_2$. DCC reacts with the carboxyl group, forming an amine-reactive O-acylisourea intermediate. (3) The intermediate is then reacted with an amine on the molecule of interest, connecting it to the porous silicon surface via a stable amide bond.

glycol (PEG) [112]. A standard method to place a PEG linker on a modified porous silicon surface is shown in Figure 6.9 [113, 114]. This reaction provides a good example of the utility of the non-aqueous DCC coupling reagent of Figure 6.8. A short-chain PEG linker yields a hydrophilic surface that is capable of passing biomolecules into or out of the pores without binding them [115]. The distal end of the PEG linker can be modified to allow coupling of other species, such as drugs, cleavable linkers, or targeting groups [113, 115].

6.5.3
Biomodification of "Hydrocarbonized" Porous Silicon

Salonen's hydrocarbonization chemistry offers an attractive, stable surface for biosensor applications. A biosensor needs to retain chemical stability in

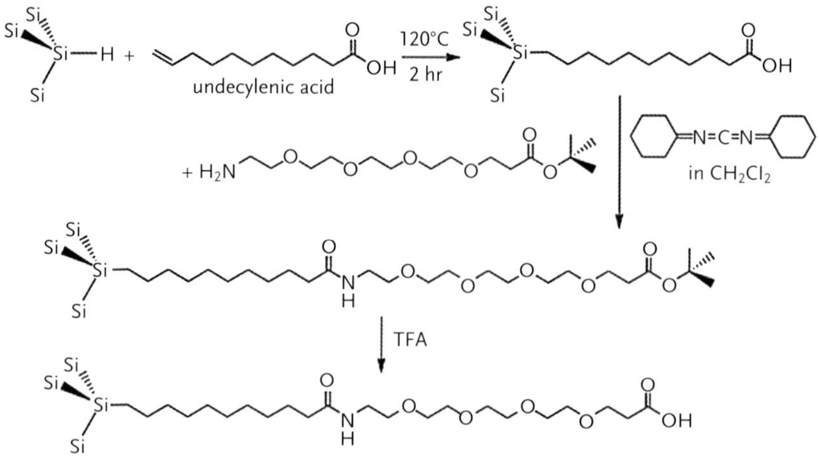

Figure 6.9 Adding a "non-stick coating" to porous silicon. The short-chain PEG linker yields a hydrophilic surface that minimizes non-specific binding of proteins to the porous silicon surface. (adapted from reference [113]). The reagent "TFA" is trifluoroacetic acid.

physiologically relevant aqueous media, and the surface chemistry must allow permeation of the aqueous solutions while also allowing attachment of specific biological capture probes. The radical coupling method of Iijima [116] to functionalize the variety of C–C and C–H bonds present on the surface of carbon fibers or diamond-like carbon has been found to work well on hydrocarbonized porous silicon (Figure 6.10) [103]. In this process, the relatively mild radical initiator, benzoyl peroxide, is used to generate surface radicals (Equation 6.21). This initiator also decarboxylates a linker molecule, sebacic acid, that is added to the reaction mixture immediately after the initiator is added (Equation 6.22). The chemistry introduces a stable carboxylic acid functionality to the surface that is quite hydrophilic, as seen in the contact angle images of Figure 6.10. The surface can then be activated to allow covalent attachment of specific sensing molecules using the methods discussed above.

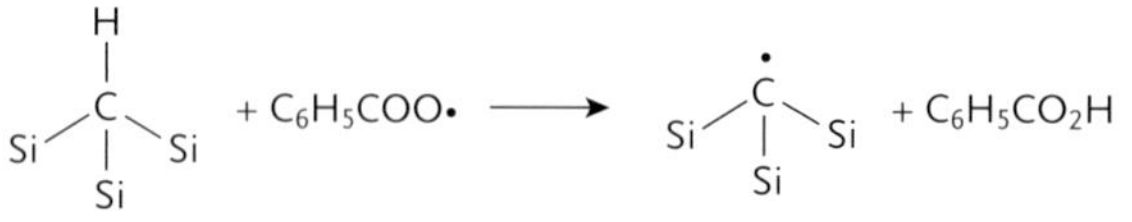

$$\text{Si-CH(H)-Si} + C_6H_5COO\bullet \longrightarrow \text{Si-C}\bullet\text{-Si} + C_6H_5CO_2H \quad (6.21)$$

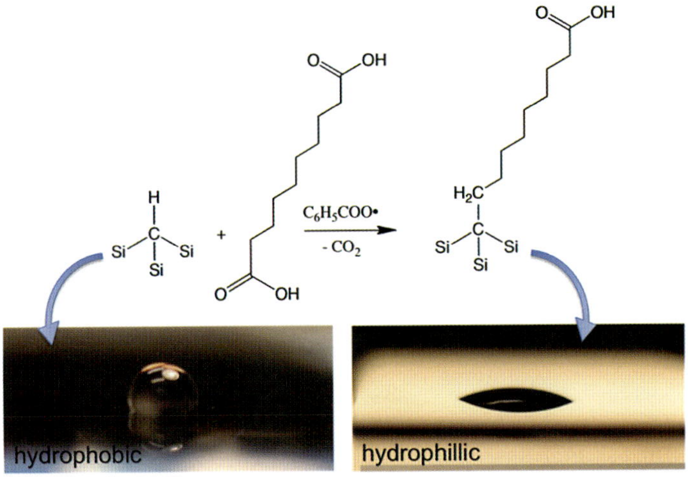

Figure 6.10
Modification of "hydrocarbonized" porous silicon using a radical coupling reaction. This stable, hydrophilic surface chemistry allows further functionalization using standard bioconjugate chemistry methods, via the easily modified carboxyl species. The surface is prepared by radical coupling of the di-carboxylic acid sebacic acid to a thermally carbonized porous silicon thin film. The two photographs show the effect on water contact angle upon modification. Image adapted from reference ([111]).

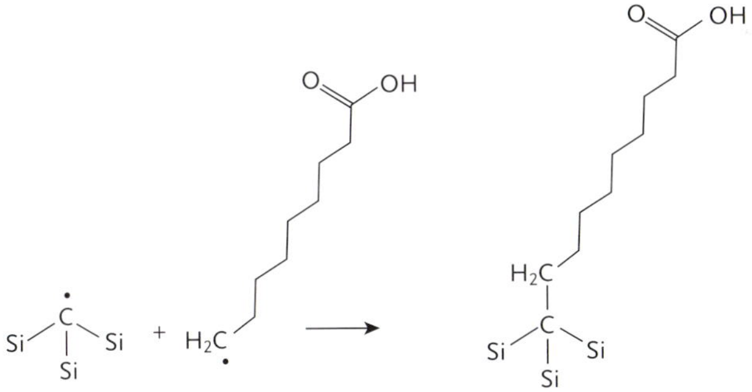

$$\tag{6.22}$$

6.5.4
Silanol-Based Coupling to Oxidized Porous Silicon Surfaces

Oxidized porous silicon can be modified using conventional silanol chemistries [35, 113, 117]. A standard coupling reaction is shown in Equation 6.23, using the reagent 3-aminopropyltriethoxysilane (APTES). Because it places the easily conjugated –NH$_2$ group on the surface, APTES is very commonly used to attach proteins, DNA, and many other molecules to silica surfaces [112, 118].

$$(6.23)$$

If one wishes to modify the interior of a microporous sample, monoalkoxy-dimethylsilanes (RO–Si(Me)$_2$–R') can be more effective than trialkoxysilanes ((RO)$_3$Si–R') as surface linkers [35, 117], because the monoalkoxysilane reagents possess only a single Si–O bond with which they link to a surface. An example is given in Equation 6.24 using the silanol 3-aminopropyldimethylethoxysilane (APDMES). The lack of additional Si–O bonds on the molecule eliminates the possibility of undesirable cross-linking reactions between other silanols. With trialkoxysilane reagents like APTES, these cross-linking reactions can produce large oligomers that clog micropores and limit the effective surface coverage.

$$(6.24)$$

One of the key design rules for *in vivo* materials is that they dissolve or are excreted after they have performed their intended function. For *in vivo* drug delivery and *in vivo* imaging, the Si–C chemistries of porous silicon can be too stable. This is particularly true of the pyrolysis-derived (carbonized and hydrocarbonized) materials. For situations where the material must eventually dissolve, chemistries involving Si–O bonds represent an attractive alternative. The time needed for highly porous SiO$_2$ to dissolve in aqueous media is consistent with many short-term drug delivery applications – typically 20 min to a few hours. In addition, a porous SiO$_2$ sample that contains no additional surface chemistries is less likely to produce toxic side effects or to induce an antigenic response. If a longer-lived oxide matrix is desired, silicon oxides formed at higher temperatures (>700 °C) are significantly more stable in aqueous media than those formed at lower temperatures or by ozone oxidation [119]. To obtain effective coupling of the alkoxysilane to these surfaces (Equations 6.23 or 6.24), the thermally oxidized samples should be soaked in a 1 M aqueous HCl solution for a few minutes, rinsed with water and then with ethanol, and briefly dried prior to exposure to the alkoxysilane. The acid rinse hydrates and protonates the Si–O surface, allowing more effective coupling of the alkoxysilane species. A typical procedure to attach the alkoxysilane species: Immerse the oxidized, hydroxylated sample in a 10% alkoxysilane solution (in heptane) under an inert atmosphere. Heat to 80 °C for 5 h, cool, remove the sample, and rinse with heptane, dichloromethane, and finally ethanol.

6.6 Chemical Modification in Tandem with Etching

The stability of Si–C bonds in the presence of HF enables the stratification of chemical functionalities within a porous silicon nanostructure [120]. To create such stacks, electrochemical etching and chemical modification steps are performed in tandem, as shown in Figure 6.11. In that example, a sample is prepared that contains both a hydrophilic layer and a hydrophobic

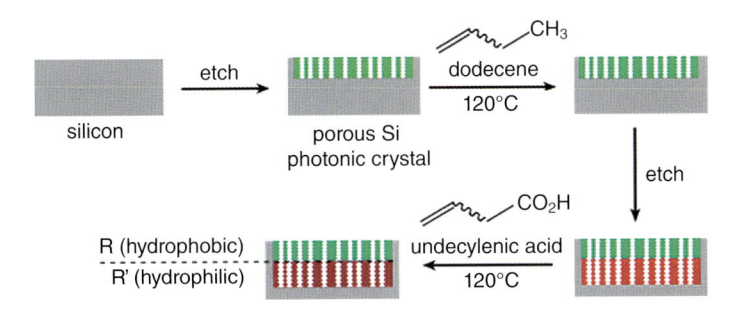

Figure 6.11
Stratification of chemistries in porous silicon. A silicon wafer is electrochemically etched and then chemically modified via hydrosilylation (in this example, with 1-dodecene). The sample is placed back in an etch cell, and a second porous layer is then etched beneath the first stack, and is subsequently modified with undecylenic acid. The end result is a double layer with a top hydrophobic layer and a bottom hydrophilic layer. After reference ([121]).

layer. The chemistry relies on the fact that the Si–C bonds formed in the hydrosilylation step are not removed by the HF-containing electrolyte.

The tandem etch/modify procedure has been used to create a variety of bifunctional nanostructures from porous silicon [33, 120–126]. Although it has not yet been reported, the procedure should allow any number of layers with distinct chemistries and pore morphologies. Particles made from the bilayer structures show amphiphilic properties [33, 120].

6.7
Metallization Reactions of Porous Silicon

In the sections discussing oxidation reactions, we saw that porous silicon is a good reducing agent due to the presence of Si–H species on the surface and elemental silicon in the skeleton. The reduction potential of either of these species is sufficient to reduce many organic molecules. For example, porous silicon has been found to interfere with the MTT cell viability assay by directly reducing MTT to formazan [127], and it can reduce benzoquinone to hydroquinone [27]. Therefore, it is not surprising that porous silicon can reduce many metal ions down to their elemental state [128–132]. The basic mechanism was worked out by Ogata and his coworkers, who demonstrated electroless deposition of various noble and coinage metals in and on porous silicon via galvanic displacement [131–134]. Once a small metallic nucleation center is formed, it will promote further reduction of the metal ions still in solution, in a manner similar to the development of silver grains in the photographic process. Two challenges with the use of as-formed porous silicon are that the metallization reaction is very rapid, and the surface hydrides make the inner pore walls very hydrophobic and impermeable to water unless a surfactant is present. Metals tend to be deposited on the porous silicon surface in patches, rather than as a conformal layer [128–132, 135, 136]. The patchy morphology has been explained in terms of a localized deposition mechanism by Ogata [131]. In Ogata's local cell model, the oxidation half-reaction corresponds to Si and Si–H species oxidizing to SiO_2, idealized as Equation 6.25. This anodic half-reaction is spatially separated from the cathodic metal deposition process (Equation 6.26), which occurs preferentially on growing metallic nuclei. Electrons transfer between the two redox-active regions through the conductive porous silicon skeleton.

$$Si + 2H_2O \rightarrow SiO_2 + 4H^+ + 4e^- \qquad (6.25)$$

$$M^{n+} + ne^- \rightarrow M \qquad (6.26)$$

The reduction kinetics of Equation 6.26 can be slowed by various additives to the solution or by various surface treatments. The reaction is also affected by temperature and applied bias; adjustment of these parameters determines whether the pores are filled completely or capped at the mouth [133, 137].

References

1. Dancil, K.-P.S., Greiner, D.P., and Sailor, M.J. (1999) A porous silicon optical biosensor: detection of reversible binding of IgG to a protein A-modified surface. *J. Am. Chem. Soc.*, **121** (34), 7925–7930.
2. Anglin, E.J., Cheng, L., Freeman, W.R., and Sailor, M.J. (2008) Porous silicon in drug delivery devices and materials. *Adv. Drug Deliv. Rev.*, **60** (11), 1266–1277.
3. Chiappini, C., Tasciotti, E., Fakhoury, J.R., Fine, D., Pullan, L., Wang, Y.C., Fu, L.F., Liu, X.W., and Ferrari, M. (2010) Tailored porous silicon microparticles: fabrication and properties. *Chem. Phys. Chem.*, **11** (5), 1029–1035.
4. Salonen, J., Kaukonen, A.M., Hirvonen, J., and Lehto, V.-P. (2008) Mesoporous silicon in drug delivery applications. *J. Pharm. Sci.*, **97** (2), 632–653.
5. Bjorklund, R.B., Zangooie, S., and Arwin, H. (1997) Adsorption of Surfactants in porous silicon films. *Langmuir*, **13** (6), 1440–1445.
6. Canaria, C.A., Huang, M., Cho, Y., Heinrich, J.L., Lee, L.I., Shane, M.J., Smith, R.C., Sailor, M.J., and Miskelly, G.M. (2002) The effect of surfactants on the reactivity and photophysics of luminescent nanocrystalline porous silicon. *Adv. Funct. Mater.*, **12** (8), 495–500.
7. Ogata, Y., Niki, H., Sakka, T., and Iwasaki, M. (1995) Oxidation of porous silicon under water-vapor environment. *J. Electrochem. Soc.*, **142** (5), 1595–1601.
8. Pacholski, C., Sartor, M., Sailor, M.J., Cunin, F., and Miskelly, G.M. (2005) Biosensing using porous silicon double-layer interferometers: reflective interferometric Fourier transform spectroscopy. *J. Am. Chem. Soc.*, **127** (33), 11636–11645.
9. Petrova-Koch, V., Muschik, T., Kux, A., Meyer, B.K., Koch, F., and Lehmann, V. (1992) Rapid thermal oxidized porous Si- the superior photoluminescent Si. *Appl. Phys. Lett.*, **61** (8), 943–945.
10. Song, J.H., and Sailor, M.J. (1998) Dimethyl sulfoxide as a mild oxidizing agent for porous silicon and its effect on photoluminescence. *Inorg. Chem.*, **37** (13), 3355–3360.
11. Mattei, G., Alieva, E.V., Petrov, J.E., and Yakovlev, V.A. (2000) Quick oxidation of porous silicon in presence of pyridine vapor. *Phys. Status Solidi A-Appl. Res.*, **182** (1), 139–143.
12. Unagami, T. (1997) Intrinsic stress in porous silicon layers formed by anodization in HF solution. *J. Electrochem. Soc.*, **144** (5), 1835–1838.
13. Anglin, E.J., Schwartz, M.P., Ng, V.P., Perelman, L.A., and Sailor, M.J. (2004) Engineering the chemistry and nanostructure of porous silicon Fabry-Pérot films for loading and release of a steroid. *Langmuir*, **20** (25), 11264–11269.
14. Kolasinski, K.W. (2005) Silicon nanostructures from electroless electrochemical etching. *Curr. Opin. Solid State Mater. Sci.*, **9** (1–2), 73–83.
15. Ghulinyan, M., Gelloz, B., Ohta, T., Pavesi, L., Lockwood, D.J., and Koshida, N. (2008) Stabilized porous silicon optical superlattices with controlled surface passivation. *Appl. Phys. Lett.*, **93** (6), 061113.
16. Salhi, B., Gelloz, B., Koshida, N., Patriarche, G., and Boukherroub, R. (2007) Synthesis and

photoluminescence properties of silicon nanowires treated by high-pressure water vapor annealing. *Phys. Status Solidi A-Appl. Mater*, **204** (5), 1302–1306.
17. Gelloz, B., Mentek, R., and Koshida, N. (2009) Specific blue light emission from nanocrystalline porous Si treated by high-pressure water vapor annealing. *Jpn. J. Appl. Phys.*, **48** (4), 04c119.
18. Gelloz, B., and Koshida, N. (2009) Long-lived blue phosphorescence of oxidized and annealed nanocrystalline silicon. *Appl. Phys. Lett.*, **94** (20), 201903.
19. Gupta, P., Dillon, A.C., Bracker, A.S., and George, S.M. (1991) FTIR studies of H_2O and D_2O decomposition on porous silicon. *Surf. Sci.*, **245**, 360–372.
20. Lehmann, V. (2002) *Electrochemistry of Silicon*, Wiley-VCH Verlag GmbH, Weinheim, Germany, pp. 51–75.
21. Sweryda-Krawiek, B., Chandler-Henderson, R.R., Coffer, J.L., Rho, Y.G., and Pinizzotto, R.F. (1996) A comparison of porous silicon and silicon nanocrystallite photoluminescence quenching with amines. *J. Phys. Chem.*, **100**, 13776–13780.
22. Sailor, M.J., and Wu, E.C. (2009) Photoluminescence-based sensing with porous silicon films, microparticles, and nanoparticles. *Adv. Funct. Mater.*, **19** (20), 3195–3208.
23. Létant, S.E., Content, S., Tan, T.T., Zenhausern, F., and Sailor, M.J. (2000) Integration of porous silicon chips in an electronic artificial nose. *Sens. Actuators B*, **69** (1–2), 193–198.
24. Canham, L.T., Newey, J.P., Reeves, C.L., Houlton, M.R., Loni, A., Simmons, A.J., and Cox, T.I. (1996) The effects of DC electric currents on the in-vitro calcification of bioactive Si wafers. *Adv. Mater.*, **8** (10), 847–849.
25. Canham, L.T. (1995) Bioactive silicon structure fabrication through nanoetching techniques. *Adv. Mater.*, **7** (12), 1033–1037.
26. Canham, L.T., Reeves, C.L., Loni, A., Houlton, M.R., Newey, J.P., Simons, A.J., and Cox, T.I. (1997) Calcium phosphate nucleation on porous silicon: factors influencing kinetics in acellular simulated body fluids. *Thin Sol. Films*, **297** (1–2), 304–307.
27. Harper, T.F., and Sailor, M.J. (1997) Using porous silicon as a hydrogenating agent: derivatization of the surface of luminescent nanocrystalline silicon with benzoquinone. *J. Am. Chem. Soc.*, **119** (29), 6943–6944.
28. Boukherroub, R., Morin, S., Sharpe, P., Wayner, D.D.M., and Allongue, P. (2000) Insights into the formation mechanisms of Si-OR monolayers from the thermal reactions of alcohols and aldehydes with Si(111)-H. *Langmuir*, **16** (19), 7429–7434.
29. Kopping, B., Chatgilialoglu, C., Zehnder, M., and Geise, B. (1992) Tris(trimethylsilyl)silane: an efficient hydrosilylating agent of alkenes and alkynes. *J. Org. Chem.*, **57** (14), 3994–4000.
30. Gu, L., Park, J.-H., Duong, K.H., Ruoslahti, E., and Sailor, M.J. (2010) Magnetic luminescent porous silicon microparticles for localized delivery of molecular drug payloads. *Small*, **6**, 2546–2552.
31. Granitzer, P., Rumpf, K., Roca, A.G., Morales, M.P., Poelt, P., and Albu, M. (2010) Magnetite nanoparticles embedded in biodegradable porous silicon. *J. Magn. Magn. Mater.*, **322**, 1343–1346.
32. Thomas, J.C., Pacholski, C., and Sailor, M.J. (2006) Delivery of nanogram payloads using magnetic porous silicon microcarriers. *Lab Chip*, **6** (6), 782–787.
33. Dorvee, J.R., Derfus, A.M., Bhatia, S.N., and Sailor, M.J. (2004) Manipulation of liquid droplets using amphiphilic, magnetic 1-D photonic crystal chaperones. *Nat. Mater.*, **3**, 896–899.
34. Kelly, T.L., Gao, T., and Sailor, M.J. (2011) Carbon and carbon/silicon

composites templated in microporous silicon rugate filters for the adsorption and detection of organic vapors. *Adv. Mater.*, **23**, 1776–1781.

35 Tinsley-Bown, A.M., Canham, L.T., Hollings, M., Anderson, M.H., Reeves, C.L., Cox, T.I., Nicklin, S., Squirrell, D.J., Perkins, E., Hutchinson, A., Sailor, M.J., and Wun, A. (2000) Tuning the pore size and surface chemistry of porous silicon for immunoassays. *Phys. Status Solidi A-Appl. Res.*, **182** (1), 547–553.

36 Halimaoui, A. (1997) Porous silicon formation by anodisation, in *Properties of Porous Silicon*, vol. 18 (ed. L. Canham), Institution of Engineering and Technology, London, pp. 12–22.

37 Collins, B.E., Dancil, K.-P., Abbi, G., and Sailor, M.J. (2002) Determining protein size using an electrochemically machined pore gradient in silicon. *Adv. Funct. Mater.*, **12** (3), 187–191.

38 Arwin, H., Gavutis, M., Gustafsson, J., Schultzberg, M., Zangooie, S., and Tengvall, P. (2000) Protein adsorption in thin porous silicon layers. *Phys. Status Solidi A-Appl. Res.*, **182**, 515–520.

39 Lust, S., and Levy-Clement, C. (2000) Macropore formation on medium doped p-type silicon. *Phys. Status Solidi A-Appl. Res.*, **182** (1), 17–21.

40 Ponomarev, E.A., and Levy-Clement, C. (2000) Macropore formation on p-type silicon. *J. Porous Mater.*, **7** (1–3), 51–56.

41 Harraz, F.A., El-Sheikh, S.M., Sakka, T., and Ogata, Y.H. (2008) Cylindrical pore arrays in silicon with intermediate nano-sizes: a template for nanofabrication and multilayer applications. *Electrochim. Acta*, **53** (22), 6444–6451.

42 Salem, M.S., Sailor, M.J., Fukami, K., Sakka, T., and Ogata, Y.H. (2009) Preparation and optical properties of porous silicon rugate-type multilayers with different pore sizes. *Phys. Status Solidi C*, **6** (7), 1620–1623.

43 Lehmann, V., Stengl, R., and Luigart, A. (2000) On the morphology and the electrochemical formation mechanism of mesoporous silicon. *Mater. Sci. Eng. B*, **B69–70**, 11–22.

44 Jugdaohsingh, R. (2007) Silicon and bone health. *J. Nutr. Health Aging*, **11** (2), 99–110.

45 Canham, L.T., Saunders, S.J., Heeley, P.B., Keir, A.M., and Cox, T.I. (1994) Rapid chemistry of porous silicon undergoing hydrolysis. *Adv. Mater.*, **6** (11), 865–868.

46 Jay, T., Canham, L.T., Heald, K., Reeves, C.L., and Downing, R. (2000) Autoclaving of porous silicon within a hospital environment: potential benefits and problems. *Phys. Status Solidi A-Appl. Res.*, **182** (1), 555–560.

47 Omae, K., Sakai, T., Sakurai, H., Yamazaki, K., Shibata, T., Mori, K., Kudo, M., Kanoh, H., and Tati, M. (1992) Acute and subacute inhalation toxicity of silane 1000 ppm in mice. *Arch. Toxicol.*, **66** (10), 750–753.

48 Takebayashi, T. (1993) Acute inhalation toxicity of high-concentrations of silane in male icr mice. *Arch. Toxicol.*, **67** (1), 55–60.

49 Kawanabe, K., Yamamuro, T., Kotani, S., and Nakamura, T. (1992) Acute nephrotoxicity as an adverse effect after intraperitoneal injection of massive amounts of bioactive ceramic powders in mice and rats. *J. Biomed. Mater. Res.*, **26** (2), 209–219.

50 Gorustovich, A.A., Monserrat, A.J., Guglielmotti, M.B., and Cabrini, R.L. (2007) Effects of intraosseous implantation of silica-based bioactive glass particles on rat kidney under experimental renal failure. *J. Biomater. Appl.*, **21** (4), 431–442.

51 Lai, W., Garino, J., Flaitz, C., and Ducheyne, P. (2005) Excretion of resorption products from bioactive

glass implanted in rabbit muscle. *J. Biomed. Mater. Res. A,* **75A** (2), 398–407.

52 Canham, L.T., Reeves, C.L., Newey, J.P., Houlton, M.R., Cox, T.I., Buriak, J.M., and Stewart, M.P. (1999) Derivatized mesoporous silicon with dramatically improved stability in simulated human blood plasma. *Adv. Mater.,* **11** (18), 1505–1507.

53 Canham, L.T., Stewart, M.P., Buriak, J.M., Reeves, C.L., Anderson, M., Squire, E.K., Allcock, P., and Snow, P.A. (2000) Derivatized porous silicon mirrors: implantable optical components with slow resorbability. *Phys. Status Solidi A-Appl. Res.,* **182** (1), 521–525.

54 Foraker, A.B., Walczak, R.J., Cohen, M.H., Boiarski, T.A., Grove, C.F., and Swaan, P.W. (2003) Microfabricated porous silicon particles enhance paracellular delivery of insulin across intestinal Caco-2 cell monolayers. *Pharm. Res.,* **20** (1), 110–116.

55 Bayliss, S.C., Harris, P.J., Buckberry, L.D., and Rousseau, C. (1997) Phosphate and cell growth on nanostructured semiconductors. *J. Mater. Sci. Lett.,* **16**, 737–740.

56 Bayliss, S.C., Buckberry, L.D., Fletcher, I., and Tobin, M.J. (1999) The culture of neurons on silicon. *Sens. Actuators A,* **74** (1–3), 139–142.

57 Bayliss, S.C., Heald, R., Fletcher, D.I., and Buckberry, L.D. (1999) The culture of mammalian cells on nanostructured silicon. *Adv. Mater.,* **11** (4), 318–321.

58 Bayliss, S.C., Buckberry, L.D., Harris, P.J., and Tobin, M. (2000) Nature of the silicon-animal cell interface. *J. Porous Mater.,* **7** (1–3), 191–195.

59 Mayne, A.H., Bayliss, S.C., Barr, P., Tobin, M., and Buckberry, L.D. (2000) Biologically interfaced porous silicon devices. *Phys. Status Solidi A-Appl. Res.,* **182** (1), 505–513.

60 Chin, V., Collins, B.E., Sailor, M.J., and Bhatia, S.N. (2001) Compatibility of primary hepatocytes with oxidized nanoporous silicon. *Adv. Mater.,* **13** (24), 1877–1880.

61 Linford, M.R., and Chidsey, C.E.D. (1993) Alkyl monolayers covalently attached to silicon surfaces. *J. Am. Chem. Soc.,* **115**, 12631–12632.

62 Buriak, J.M. (1999) Organometallic chemistry on silicon surfaces: formation of functional monolayers bound through Si-C bonds. *Chem. Commun.,* (12), 1051–1060.

63 Buriak, J.M. (2002) Organometallic chemistry on silicon and germanium surfaces. *Chem. Rev.,* **102** (5), 1272–1308.

64 Buriak, J.M., and Allen, M.J. (1998) Lewis acid mediated functionalization of porous silicon with substituted alkenes and alkynes. *J. Am. Chem. Soc.,* **120** (6), 1339–1340.

65 Boukherroub, R., Morin, S., Wayner, D.D.M., and Lockwood, D.J. (2000) Thermal route for chemical modification and photoluminescence stabilization of porous silicon. *Phys. Status Solidi A-Appl. Res.,* **182** (1), 117–121.

66 Boukherroub, R., Morin, S., Wayner, D.D.M., Bensebaa, F., Sproule, G.I., Baribeau, J.M., and Lockwood, D.J. (2001) Ideal passivation of luminescent porous silicon by thermal, noncatalytic reaction with alkenes and aldehydes. *Chem. Mater.,* **13** (6), 2002–2011.

67 Boukherroub, R., Petit, A., Loupy, A., Chazalviel, J.N., and Ozanam, F. (2003) Microwave-assisted chemical functionalization of hydrogen-terminated porous silicon surfaces. *J. Phys. Chem. B,* **107** (48), 13459–13462.

68 Boukherroub, R., Wojtyk, J.T.C., Wayner, D.D.M., and Lockwood, D.J. (2002) Thermal hydrosilylation of undecylenic acid with porous silicon. *J. Electrochem. Soc.,* **149** (2), 59–63.

69 Mattei, G., and Valentini, V. (2003) *In situ* functionalization of porous silicon during the electrochemical

formation process in ethanoic hydrofluoric acid solution. *J. Am. Chem. Soc.*, **125** (32), 9608–9609.

70 Sieval, A.B., Demirel, A.L., Nissink, J.W.M., Linford, M.R., van der Maas, J.H., de Jeu, W.H., Zuilhof, H., and Sudholter, E.J.R. (1998) Highly stable Si-C linked functionalized monolayers on the silicon (100) surface. *Langmuir*, **14** (7), 1759–1768.

71 Sieval, A.B., Linke, R., Zuilhof, H., and Sudholter, E.J.R. (2000) High-quality alkyl monolayers on silicon surfaces. *Adv. Mater.*, **12** (19), 1457–1460.

72 Stewart, M.P., and Buriak, J.M. (1998) Photopatterned hydrosilylation on porous silicon. *Angew. Chem. Int. Ed. Engl.*, **37** (23), 3257–3260.

73 Stewart, M.P., and Buriak, J.M. (2001) Exciton-mediated hydrosilylation on photoluminescent nanocrystalline silicon. *J. Am. Chem. Soc.*, **123**, 7821–7830.

74 Niederhauser, T.L., Lua, Y.Y., Sun, Y., Jiang, G.L., Strossman, G.S., Pianetta, P., and Linford, M.R. (2002) Formation of (functionalized) monolayers and simultaneous surface patterning by scribing silicon in the presence of alkyl halides. *Chem. Mater.*, **14** (1), 27–29.

75 Shriver, D.F., and Drezdzon, M.A. (1986) *The Manipulation of Air-Sensitive Compounds*, 2nd edn, John Wiley & Sons, Inc., New York, pp. 7–44.

76 Ghicov, A., and Schmuki, P. (2009) Self-ordering electrochemistry: a review on growth and functionality of TiO_2 nanotubes and other self-aligned MO_x structures. *Chem. Commun.*, (20), 2791–2808.

77 Lindsay, S., Vazquez, T., Egatz-Gomez, A., Loyprasert, S., Garcia, A.A., and Wang, J. (2007) Discrete microfluidics with electrochemical detection. *Analyst*, **132** (5), 412–416.

78 Coffinier, Y., Janel, S., Addad, A., Blossey, R., Gengembre, L., Payen, E., and Boukherroub, R. (2007) Preparation of superhydrophobic silicon oxide nanowire surfaces. *Langmuir*, **23** (4), 1608–1611.

79 Singh, S., Houston, J., van Swol, F., and Brinker, C.J. (2006) Superhydrophobicity–drying transition of confined water. *Nature*, **442** (7102), 526–526.

80 Krupenkin, T.N., Taylor, J.A., Schneider, T.M., and Yang, S. (2004) From rolling ball to complete wetting: the dynamic tuning of liquids on nanostructured surfaces. *Langmuir*, **20** (10), 3824–3827.

81 Linford, M.R., Fenter, P., Eisenberger, P.M., and Chidsey, C.E.D. (1995) Alkyl monolayers on silicon prepared from 1-alkenes and hydrogen-terminated silicon. *J. Am. Chem. Soc.*, **117**, 3145–3155.

82 Terry, J., Linford, M.R., Wigren, C., Cao, R., Pianetta, P., and Chidsey, C.E.D. (1997) Determination of the bonding of alkyl monolayers to the Si(111) surface using chemical-shift, scanned-energy photoelectron diffraction. *Appl. Phys. Lett.*, **71** (8), 1056–1058.

83 Terry, J., Linford, M.R., Wigren, C., Cao, R.Y., Pianetta, P., and Chidsey, C.E.D. (1999) Alkyl-terminated Si(111) surfaces: a high-resolution, core level photoelectron spectroscopy study. *J. Appl. Phys.*, **85** (1), 213–221.

84 Song, J.H., and Sailor, M.J. (1998) Functionalization of nanocrystalline porous silicon surfaces with aryllithium reagents: formation of silicon-carbon bonds by cleavage of silicon-silicon bonds. *J. Am. Chem. Soc.*, **120** (10), 2376–2381.

85 Song, J.H., and Sailor, M.J. (1999) Chemical modification of crystalline porous silicon surfaces. *Comments Inorg. Chem.*, **21** (1–3), 69–84.

86 Bansal, A., and Lewis, N.S. (1998) Stabilization of Si photoanodes in aqueous electrolytes through surface alkylation. *J. Phys. Chem. B*, **102** (21), 4058–4060.

87 Bansal, A., Li, X., Lauermann, I., Lewis, N.S., Yi, S.I., and Weinberg, W.H. (1996) Alkylation of Si

88 Dubois, T., Ozanam, F., and Chazalviel, J.-N. (1997) Stabilization of the porous silicon surface by grafting of organic groups: direct electrochemical methylation. *Proc. Electrochem. Soc.*, 97–7, 296–311.

surfaces using a two-step halogenation/grignard route. *J. Am. Chem. Soc.*, **118**, 7225–7226.

89 Gurtner, C., Wun, A.W., and Sailor, M.J. (1999) Surface modification of porous silicon by electrochemical reduction of organo halides. *Angew. Chem. Int. Ed.*, **38** (13/14), 1966–1968.

90 Thompson, C.M., Nieuwoudt, M., Ruminski, A.M., Sailor, M.J., and Miskelly, G.M. (2010) Electrochemical preparation of pore wall modification gradients across thin porous silicon layers. *Langmuir*, **26** (10), 7598–7603.

91 Orosco, M.M., Pacholski, C., Miskelly, G.M., and Sailor, M.J. (2006) Protein-coated porous silicon photonic crystals for amplified optical detection of protease activity. *Adv. Mater.*, **18**, 1393–1396.

92 Lees, I.N., Lin, H., Canaria, C.A., Gurtner, C., Sailor, M.J., and Miskelly, G.M. (2003) Chemical stability of porous silicon surfaces electrochemically modified with functional alkyl species. *Langmuir*, **19**, 9812–9817.

93 Salonen, J., Lehto, V.P., Bjorkqvist, M., Laine, E., and Niinisto, L. (2000) Studies of thermally-carbonized porous silicon surfaces. *Phys. Status Solidi A-Appl. Res.*, **182** (1), 123–126.

94 Salonen, J., Laine, E., and Niinisto, L. (2002) Thermal carbonization of porous silicon surface by acetylene. *J. Appl. Phys.*, **91** (1), 456–461.

95 Bjorkqvist, M., Salonen, J., Laine, E., and Niinisto, L. (2003) Comparison of stabilizing treatments on porous silicon for sensor applications. *Phys. Status Solidi A-Appl. Res.*, **197** (2), 374–377.

96 Lehto, V.P., Vaha-Heikkila, K., Paski, J., and Salonen, J. (2005) Use of thermoanalytical methods in quantification of drug load in mesoporous silicon microparticles. *J. Therm. Anal. Calorim.*, **80** (2), 393–397.

97 Salonen, J., Laitinen, L., Kaukonen, A.M., Tuura, J., Bjorkqvist, M., Heikkila, T., Vaha-Heikkila, K., Hirvonen, J., and Lehto, V.P. (2005) Mesoporous silicon microparticles for oral drug delivery: loading and release of five model drugs. *J. Control. Release*, **108** (2–3), 362–374.

98 Salonen, J., Paski, J., Vaha-Heikkila, K., Heikkila, T., Bjorkqvist, M., and Lehto, V.P. (2005) Determination of drug load in porous silicon microparticles by calorimetry. *Phys. Status Solidi A-Appl. Mat.*, **202** (8), 1629–1633.

99 Limnell, T., Riikonen, J., Salonen, J., Kaukonen, A.M., Laitinen, L., Hirvonen, J., and Lehto, V.P. (2006) The effect of different surface treatment and pore size on the dissolution of ibuprofen from mesoporous silicon particles. *Eur. J. Pharm. Sci.*, **28**, S34–S34.

100 Heikkila, T., Salonen, J., Tuura, J., Kumar, N., Salmi, T., Murzin, D.Y., Hamdy, M.S., Mul, G., Laitinen, L., Kaukonen, A.M., Hirvonen, J., and Lehto, V.P. (2007) Evaluation of mesoporous TCPSi, MCM-41, SBA-15, and TUD-1 materials as API carriers for oral drug delivery. *Drug Deliv.*, **15** (6), 337–347.

101 Kaukonen, A.M., Laitinen, L., Salonen, J., Tuura, J., Heikkila, T., Limnell, T., Hirvonen, J., and Lehto, V.P. (2007) Enhanced *in vitro* permeation of furosemide loaded into thermally carbonized mesoporous silicon (TCPSi) microparticles. *Eur. J. Pharm. Biopharm.*, **66** (3), 348–356.

102 Salonen, J., Bjorkqvist, M., Laine, E., and Niinisto, L. (2004) Stabilization of porous silicon surface by thermal decomposition of acetylene. *Appl. Surf. Sci.*, **225** (1–4), 389–394.

103 Ruminski, A.M., King, B.H., Salonen, J., Snyder, J.L., and Sailor, M.J. (2010) Porous silicon-based

optical microsensors for volatile organic analytes: effect of surface chemistry on stability and specificity. *Adv. Funct. Mater.*, **20**, 2874–2883.

104 Jalkanen, T., Torres-Costa, V., Salonen, J., Björkqvist, M., Mäkilä, E., Martínez-Duart, J.M., and Lehto, V.-P. (2009) Optical gas sensing properties of thermally hydrocarbonized porous silicon Bragg reflectors. *Opt. Express*, **17** (7), 5446–5456.

105 Salonen, J., Tuura, J., Bjorkqvist, M., and Lehto, V.P. (2006) Sub-ppm trace moisture detection with a simple thermally carbonized porous silicon sensor. *Sens. Actuator B-Chem.*, **114** (1), 423–426.

106 Bjorkqvist, M., Paski, J., Salonen, J., and Lehto, V.P. (2005) Temperature dependence of thermally-carbonized porous silicon humidity sensor. *Phys. Status Solidi A-Appl. Mat.*, **202** (8), 1653–1657.

107 Bjorkqvist, M., Salonen, J., Paski, J., and Laine, E. (2004) Characterization of thermally carbonized porous silicon humidity sensor. *Sens. Actuator A-Phys.*, **112** (2–3), 244–247.

108 Bjorkqvist, M., Salonen, J., and Laine, E. (2004) Humidity behavior of thermally carbonized porous silicon. *Appl. Surf. Sci.*, **222** (1–4), 269–274.

109 Hermanson, G.T. (1996) *Bioconjugate Techniques*, Academic Press, Inc., San Diego, p. 785.

110 Wu, E.C., Andrew, J.S., Cheng, L., Freeman, W.R., Pearson, L., and Sailor, M.J. (2011) Real-time monitoring of sustained drug release using the optical properties of porous silicon photonic crystal particles. *Biomaterials*, **32**, 1957–1966.

111 Sciacca, B., Alvarez, S.D., Geobaldo, F., and Sailor, M.J. (2010) Bioconjugate functionalization of thermally carbonized porous silicon using a radical coupling reaction. *Dalton Trans.*, **39**, 10847–10853.

112 Anderson, A.S., Dattelbaum, A.M., Montano, G.A., Price, D.N., Schmidt, J.G., Martinez, J.S., Grace, W.K., Grace, K.M., and Swanson, B.I. (2008) Functional PEG-modified thin films for biological detection. *Langmuir*, **24** (5), 2240–2247.

113 Schwartz, M.P., Cunin, F., Cheung, R.W., and Sailor, M.J. (2005) Chemical modification of silicon surfaces for biological applications. *Phys. Status Solidi A-Appl. Mat.*, **202** (8), 1380–1384.

114 Kilian, K.A., Bocking, T., Gaus, K., Gal, M., and Gooding, J.J. (2007) Si-C linked oligo(ethylene glycol) layers in silicon-based photonic crystals: optimization for implantable optical materials. *Biomaterials*, **28**, 3055–3062.

115 Kilian, K.A., Böcking, T., Gaus, K., Gal, M., and Gooding, J.J. (2007) Peptide-modified optical filters for detecting protease activity. *ACS Nano*, **1** (4), 355–361.

116 Iijima, M., and Kamiya, H. (2008) Surface modification of silicon carbide nanoparticles by azo radical initiators. *J. Phys. Chem. C*, **112** (31), 11786–11790.

117 Janshoff, A., Dancil, K.-P.S., Steinem, C., Greiner, D.P., Lin, V.S.-Y., Gurtner, C., Motesharei, K., Sailor, M.J., and Ghadiri, M.R. (1998) Macroporous p-type silicon Fabry-Perot layers. Fabrication, characterization, and applications in biosensing. *J. Am. Chem. Soc.*, **120** (46), 12108–12116.

118 Nijdam, A.J., Cheng, M.M.C., Geho, D.H., Fedele, R., Herrmann, P., Killian, K., Espina, V., Petricoin, E.F., Liotta, L.A., and Ferrari, M. (2007) Physicochemically modified silicon as a substrate for protein microarrays. *Biomaterials*, **28** (3), 550–558.

119 Pacholski, C., Yu, C., Miskelly, G.M., Godin, D., and Sailor, M.J. (2006) Reflective interferometric fourier transform spectroscopy: a self-compensating label-free immunosensor using double-layers

of porous SiO$_2$. *J. Am. Chem. Soc.*, **128**, 4250–4252.

120 Link, J.R., and Sailor, M.J. (2003) Smart dust: self-assembling, self-orienting photonic crystals of porous Si. *Proc. Natl. Acad. Sci. U S A*, **100** (19), 10607–10610.

121 Ruminski, A.M., Moore, M.M., and Sailor, M.J. (2008) Humidity-compensating sensor for volatile organic compounds using stacked porous silicon photonic crystals. *Adv. Funct. Mater.*, **18** (21), 3418–3426.

122 Sailor, M.J., and Link, J.R. (2005) Smart dust: nanostructured devices in a grain of sand. *Chem. Commun.*, 1375–1383.

123 Dorvee, J.R., Sailor, M.J., and Miskelly, G.M. (2008) Digital microfluidics and delivery of molecular payloads with magnetic porous silicon chaperones. *Dalton Trans.*, **6**, 721–730.

124 Kilian, K.A., Bocking, T., Lai, L.M.H., Ilyas, S., Gaus, K., Gal, M., and Gooding, J.J. (2008) Organic modification of mesoporous silicon rugate filters: the influence of nanoarchitecture on optical behaviour. *Int. J. Nanotechnol.*, **5** (2–3), 170–178.

125 Ciampi, S., Bocking, T., Kilian, K.A., Harper, J.B., and Gooding, J.J. (2008) Click chemistry in mesoporous materials: functionalization of porous silicon rugate filters. *Langmuir*, **24** (11), 5888–5892.

126 Kilian, K.A., Bocking, T., Ilyas, S., Gaus, K., Jessup, W., Gal, M., and Gooding, J.J. (2007) Forming antifouling organic multilayers on porous silicon rugate filters towards *in vivo/ex vivo* biophotonic devices. *Adv. Funct. Mater.*, **17** (15), 2884–2890.

127 Laaksonen, T., Santos, H., Vihola, H., Salonen, J., Riikonen, J., Heikkila, T., Peltonen, L., Kumar, N., Murzin, D.Y., Lehto, V.-P., and Hirvonen, J. (2007) Failure of MTT as a toxicity testing agent for mesoporous silicon microparticles. *Chem. Res. Toxicol.*, **20** (12), 1913–1918.

128 Parbukov, A.N., Beklemyshev, V.I., Gontar, V.M., Makhonin, I.I., Gavrilov, S.A., and Bayliss, S.C. (2001) The production of a novel stain-etched porous silicon, metallization of the porous surface and application in hydrocarbon sensors. *Mater. Sci. Eng. C*, **15** (1–2), 121–123.

129 Coulthard, I., and Sham, T.K. (1997) Synthesis of nanophase noble metal systems utilizing porous silicon. *Mater. Res. Soc. Symp. Proc.*, **457**, 161–165.

130 Coulthard, I., Jiang, D.-T., Lorimer, J.W., Sham, T.K., and Feng, X.-H. (1993) Reductive deposition of Pd on Porous Silicon from Aqueous Solutions of PdCl$_2$: an X-ray absorption fine structure study. *Langmuir*, **9**, 3441–3445.

131 Harraz, F.A., Tsuboi, T., Sasano, J., Sakka, T., and Ogata, Y.H. (2002) Metal depostion onto a porous silicon layer by immersion plating from aqueous and nonaqueous solutions. *J. Electrochem. Soc.*, **149** (9), C456–C463.

132 Tsuboi, T., Sakka, T., and Ogata, Y.H. (1998) Metal deposition into a porous silicon layer by immersion plating: influence of halide ions. *J. Appl. Phys.*, **83**, 4501–4506.

133 Fukami, K., Kobayashi, K., Matsumoto, T., Kawamura, Y.L., Sakka, T., and Ogata, Y.H. (2008) Electrodeposition of noble metals into ordered macropores in p-type silicon. *J. Electrochem. Soc.*, **155** (6), D443–D448.

134 Ogata, Y.H., Kobayashi, K., and Motoyama, M. (2006) Electrochemical metal deposition on silicon. *Curr. Opin. Solid State Mater. Sci.*, **10** (3–4), 163–172.

135 Lin, H., Mock, J., Smith, D., Gao, T., and Sailor, M.J. (2004) Surface-enhanced raman scattering from silver-plated porous silicon. *J. Phys. Chem. B*, **108** (31), 11654–11659.

136 Lin, H., Gao, T., Fantini, J., and Sailor, M.J. (2004) A porous silicon-palladium composite film for optical interferometric sensing of hydrogen. *Langmuir*, **20**, 5104–5108.

137 Fukami, K., Tanaka, Y., Chourou, M.L., Sakka, T., and Ogata, Y.H. (2009) Filling of mesoporous silicon with copper by electrodeposition from an aqueous solution. *Electrochim. Acta*, **54** (8), 2197–2202.

Appendix A1. Etch Cell Engineering Diagrams and Schematics

Engineering diagrams for electrochemical etch cells used to prepare porous Si. Designs for two different top pieces are provided:

The Standard Etch Cell has a top piece that uses a 12.37-mm ID O-ring seal. Although nominally this design exposes 1.20 cm^2 of silicon to the etching solution, compression of the O-ring and current shadowing effects near the O-ring edge yields a usable area that may be as little as 1.13 cm^2 of porous silicon. The 1.2 cm^2 value for the area is used for the calculations in this book. The cell holds 4.8 ml of electrolyte.

The Small Etch Cell has a top piece that uses a 5.23-mm ID O-ring seal. This cell is useful for photoelectrochemical etching of n-type silicon, in particular for samples which are back side-illuminated. The cell effectively exposes 0.21 cm^2 of silicon to the etch.

The Large Etch Cell has a top piece that uses a 33.05-mm ID O-ring seal. This cell effectively exposes 8.6 cm^2 of silicon to the etch. The cell holds 15 ml of electrolyte.

Standard or Small Etch Cell-Complete

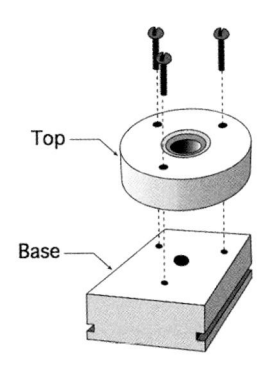

Porous Silicon in Practice: Preparation, Characterization and Applications, First Edition. Michael J. Sailor.
© 2012 Wiley-VCH Verlag GmbH & Co. KGaA. Published 2012 by Wiley-VCH Verlag GmbH & Co. KGaA.

Appendix A1. Etch Cell Engineering Diagrams and Schematics

Standard Etch Cell Top Piece

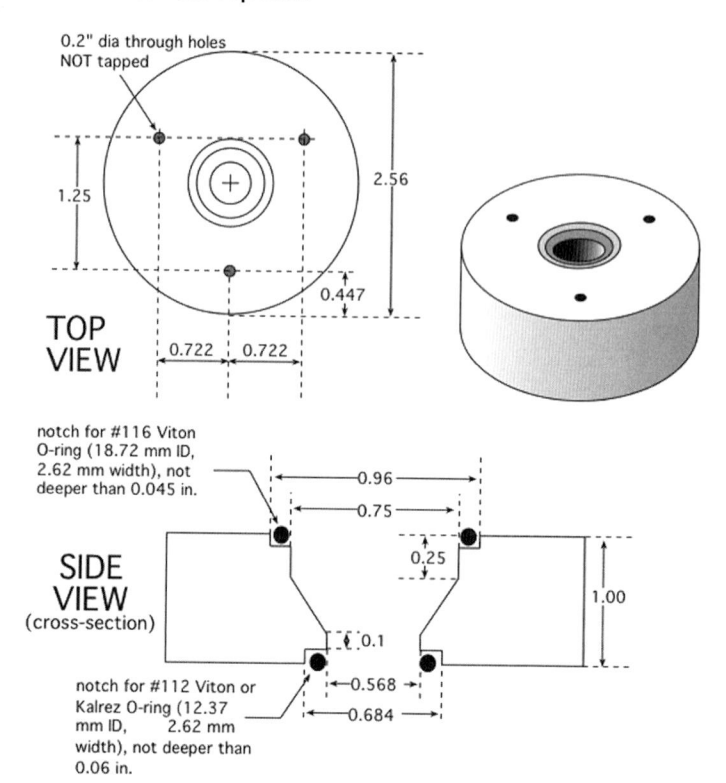

TOP VIEW

- 0.2" dia through holes NOT tapped
- 1.25
- 2.56
- 0.447
- 0.722 | 0.722

SIDE VIEW (cross-section)

- notch for #116 Viton O-ring (18.72 mm ID, 2.62 mm width), not deeper than 0.045 in.
- 0.96
- 0.75
- 0.25
- 1.00
- 0.1
- 0.568
- 0.684
- notch for #112 Viton or Kalrez O-ring (12.37 mm ID, 2.62 mm width), not deeper than 0.06 in.

Appendix A1. Etch Cell Engineering Diagrams and Schematics

Small Etch Cell Top Piece

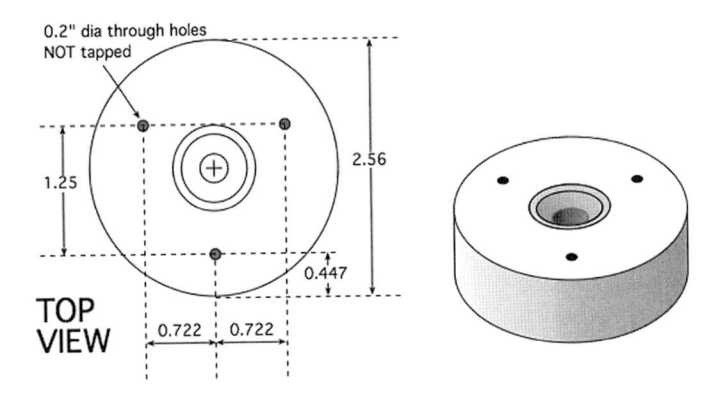

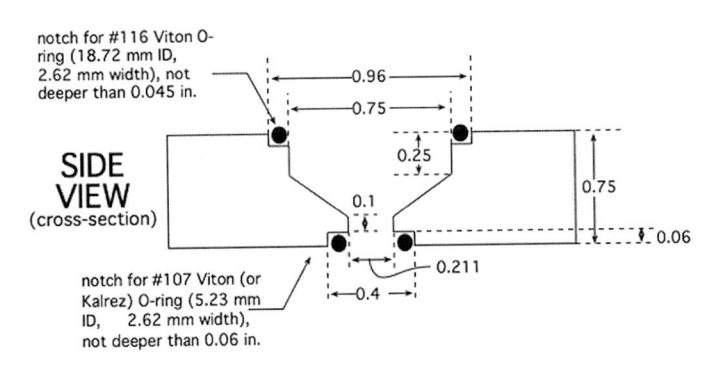

TOP VIEW

0.2" dia through holes NOT tapped

1.25
2.56
0.447
0.722 | 0.722

SIDE VIEW (cross-section)

notch for #116 Viton O-ring (18.72 mm ID, 2.62 mm width), not deeper than 0.045 in.

0.96
0.75
0.25
0.75
0.1
0.06
0.211
0.4

notch for #107 Viton (or Kalrez) O-ring (5.23 mm ID, 2.62 mm width), not deeper than 0.06 in.

Appendix A1. Etch Cell Engineering Diagrams and Schematics

Etch Cell Base (for Either Standard or Small Etch Cell)

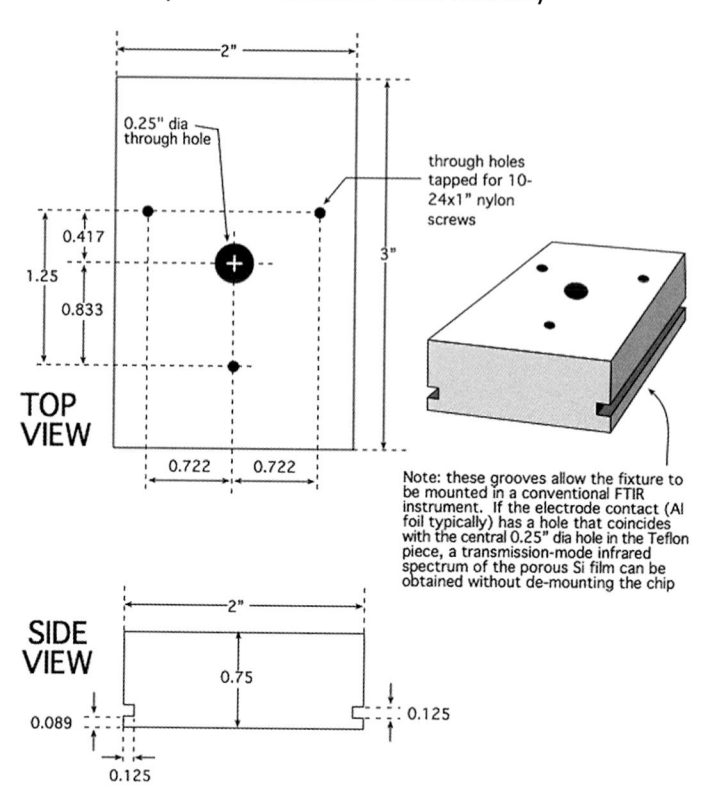

TOP VIEW — 2" × 3", 0.25" dia through hole; 0.417, 1.25, 0.833, 0.722, 0.722

through holes tapped for 10-24x1" nylon screws

Note: these grooves allow the fixture to be mounted in a conventional FTIR instrument. If the electrode contact (Al foil typically) has a hole that coincides with the central 0.25" dia hole in the Teflon piece, a transmission-mode infrared spectrum of the porous Si film can be obtained without de-mounting the chip

SIDE VIEW — 2", 0.75, 0.089, 0.125, 0.125

Large Etch Cell-Complete

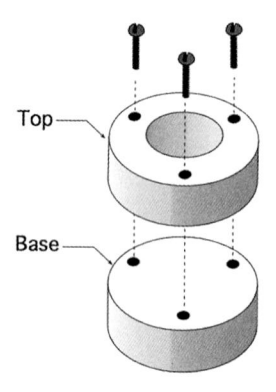

Top

Base

Appendix A1. Etch Cell Engineering Diagrams and Schematics | 233

Large Etch Cell Top Piece

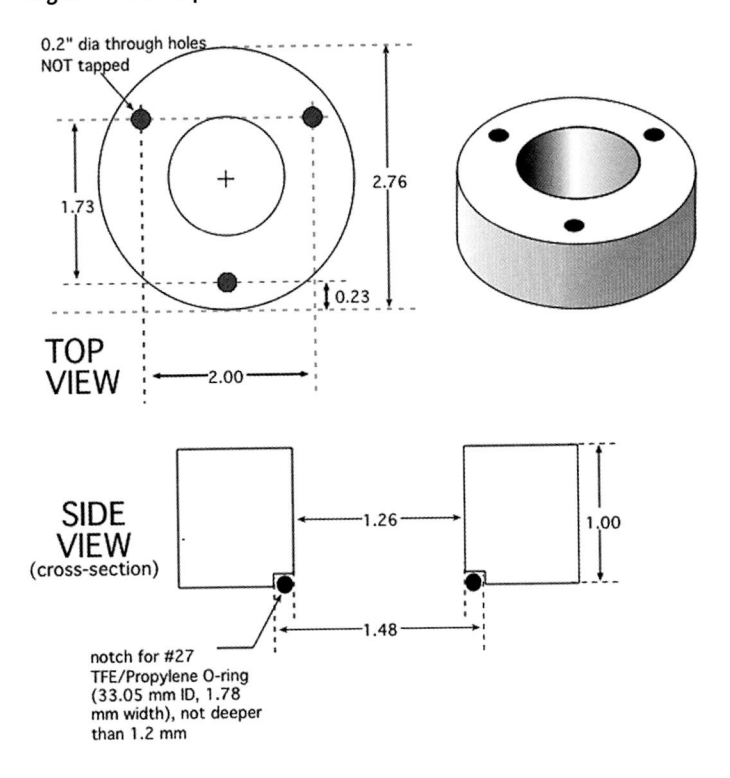

0.2" dia through holes
NOT tapped

2.76
1.73
0.23

TOP VIEW
2.00

SIDE VIEW
(cross-section)
1.26
1.00
1.48

notch for #27
TFE/Propylene O-ring
(33.05 mm ID, 1.78
mm width), not deeper
than 1.2 mm

Large Etch Cell Base

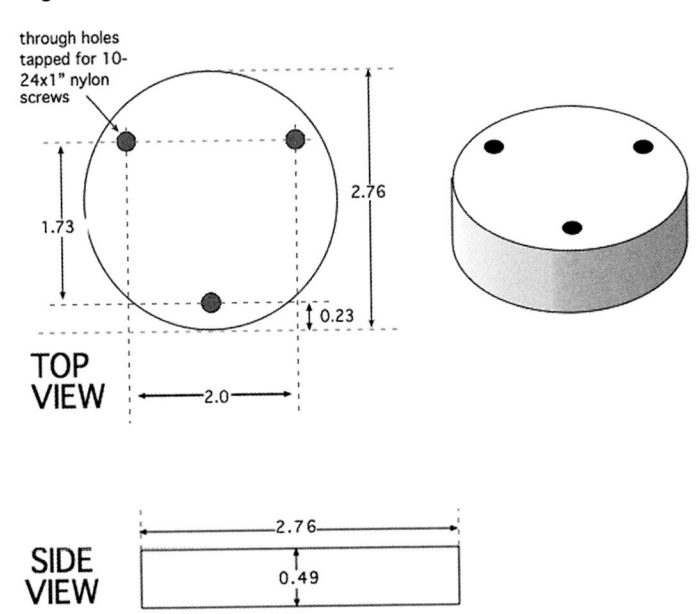

through holes
tapped for 10-24x1" nylon screws

2.76
1.73
0.23

TOP VIEW
2.0

SIDE VIEW
2.76
0.49

Appendix A2. Safety Precautions When Working with Hydrofluoric Acid

Hydrofluoric Acid Hazards

From the Mallincrodt Baker, Inc. MSDS:

Hazards Identification

Emergency Overview

POISON! DANGER! CORROSIVE. EXTREMELY HAZARDOUS LIQUID AND VAPOR. CAUSES SEVERE BURNS WHICH MAY NOT BE IMMEDIATELY PAINFUL OR VISIBLE. MAY BE FATAL IF SWALLOWED OR INHALED. LIQUID AND VAPOR CAN BURN SKIN, EYES AND RESPIRATORY TRACT. CAUSES BONE DAMAGE. REACTION WITH CERTAIN METALS GENERATES FLAMMABLE AND POTENTIALLY EXPLOSIVE HYDROGEN GAS.

J.T. Baker SAF-T-DATA(tm) Ratings:

Health Rating: 4 – Extreme (Poison)
Flammability Rating: 0 – None
Reactivity Rating: 2 – Moderate
Contact Rating: 4 – Extreme (Corrosive)
Lab Protective Equip: GOGGLES & SHIELD; LAB COAT & APRON; VENT HOOD; PROPER GLOVES
Storage Color Code: White (Corrosive)

Porous Silicon in Practice: Preparation, Characterization and Applications, First Edition.
Michael J. Sailor.
© 2012 Wiley-VCH Verlag GmbH & Co. KGaA. Published 2012 by Wiley-VCH Verlag GmbH & Co. KGaA.

Potential Health Effects:
Exposure to hydrofluoric acid can produce harmful health effects that may not be immediately apparent.

Inhalation:
Severely corrosive to the respiratory tract. May cause sore throat, coughing, labored breathing and lung congestion/inflammation.

Ingestion:
Corrosive. May cause sore throat, abdominal pain, diarrhea, vomiting, severe burns of the digestive tract, and kidney dysfunction.

Skin Contact:
Corrosive to the skin. Skin contact causes serious skin burns which may not be immediately apparent or painful. Symptoms may be delayed 8 hours or longer. The fluoride ion readily penetrates the skin causing destruction of deep tissue layers and even bone.

Eye Contact:
Corrosive to the eyes. Symptoms of redness, pain, blurred vision, and permanent eye damage may occur.

Chronic Exposure:
Intake of more than 6 mg of fluorine per day may result in fluorosis, bone and joint damage. Hypocalcemia and hypomagnesemia can occur from absorption of fluoride ion into the blood stream.

Aggravation of Pre-existing Conditions:
Persons with pre-existing skin disorders, eye problems, or impaired kidney or respiratory function may be more susceptible to the effects of this substance.

First Aid Measures for HF Contact

Inhalation:

Get medical help immediately. If patient is unconscious, give artificial respiration or use inhalator. Keep patient warm and resting, and send to hospital after first aid is complete.

Ingestion:

If swallowed, DO NOT INDUCE VOMITING. Give large quantities of water. Never give anything by mouth to an unconscious person. Get medical attention immediately.

Skin Contact:
1. Remove the victim from the contaminated area and immediately place him under a safety shower or wash him with a water hose, whichever is available.

2. Remove all contaminated clothing. Handle all HF-contaminated material with gloves made of appropriate material, such as PVC or neoprene

3. Keep washing with large amounts of water for a minimum of 15 minutes.

4. Have someone make arrangements for medical attention while you continue flushing the affected area with water.

5. If HF antidote gel (2.5% calcium gluconate in a water-soluble gel) is available, limit the washing to 5 minutes and massage the gel into the affected area.

6. Seek medical attention as soon as possible for all burns regardless of how minor they may appear initially.

ALTERNATIVES TO CALCIUM GLUCONATE GEL

Immerse the burned area in a solution of 0.2% iced aqueous Hyamine 1622 or 0.13% iced aqueous Zephiran Chloride. If immersion is not practical, towels should be soaked with one of the above solutions and used as compresses for the burn area. Ideally compresses should be changed every 2 minutes. Hyamine 1622 is a trade name for tetracaine benzethonium chloride, Merck Index Monograph 1078, a quaternary ammonium compound sold by Rohm & Haas, Philadelphia. Zephiran Chloride is a trade name for benzalkonium chloride, Merck Index Monograph 1059, also a quaternary ammonium compound, sold by Sanofi-Synthelabo Inc., New York, NY.

Eye Contact:
1. Irrigate eyes for at least 30 minutes with copious quantities of water, keeping the eyelids apart and away from eyeballs during irrigation.
2. Get competent medical attention immediately, preferably an eye specialist.
3. If a physician is not immediately available, apply one or two drops of ophthalmic anesthetic, (e.g., 0.5% Pontocaine Hydrochloride solution).
4. Do not use oily drops, ointment or HF skin burn treatments. Place ice pack on eyes until reaching emergency room.

Note to Physician

General: For burns of moderate areas, (greater than 8 square inches), ingestion and significant inhalation exposure, severe systemic effects may occur, and admission to a critical care unit should be considered. Monitor and correct for hypocalcemia, cardiac arrhythmias, hypomagnesemia and hyperkalemia. In some cases renal dialysis may be indicated.

Inhalation: Treat as chemical pneumonia. Monitor for hypocalcemia, 2.5% calcium gluconate in normal saline by nebulizer or by IPPB with 100% oxygen may decrease pulmonary damage. Bronchodilators may also be administered.

Skin: For deep skin burns or contact with concentrated HF (over 50%) solution, consider infiltration about the affected area with 5% calcium gluconate (equal parts of 10% calcium gluconate and sterile saline for injection). Burns beneath the nail may require splitting the nail and application of calcium gluconate to the exposed nail bed. For certain burns, especially of the digits, use of intra-arterial calcium gluconate may be indicated.

Eyes: Irrigation may be facilitated by use of Morgan lens or similar ocular irrigator, using 1% aqueous calcium gluconate solution (50 ml of calcium gluconate 10% in 500 ml normal saline).

AN ALTERNATIVE FIRST AID PROCEDURE

The effect of HF, that is, onset of pain, particularly in dilute solutions, may not be felt for up to 24 hours. It is important, therefore, that persons using HF have immediate access to an effective antidote, even when they are away from their work place, in order that first aid treatment can be commenced immediately.

We recommend that any person in contact with HF should carry, or have access to a tube of HF Antidote Gel at all times; ideally with one tube at the work place, one on the person and one at home.

It is imperative that any person who has been contaminated by HF should seek medical advice when the treatment by HF Antidote Gel has been applied.

HF Antidote Gel

Distributed by Pharmascience Inc.

8400 Darnley Rd. Montreal, Canada. H4T 1M4

Phone: (514) 340-1114

Fax: (514) 342-7764

U.S. (Buffalo, NY) distributor: 1-800-207-4477

You can also make your own by mixing 2.5% (by weight) calcium gluconate (Aldrich Chemicals, www.aldrich.com) in water-soluble K-Y jelly (Sav-on pharmacy)

Further Reading

Bracken, W.M., et al. (1985) Comparative effectiveness of topical treatments for hydrofluoric acid burns, University of Kansas. *J. Occup. Med.* **27**, 10.

Burke, W.J., et al. (1973) Systemic fluoride poisoning resulting from a fluoride skin burn. *J. Occup. Med.* **5**, 39.

Sprout, W.L., et al. (1993) Treatment of severe hydrofluoric acid exposures. *J. Am. Occup. Med.*, **25**, 12.

Appendix A3. Gas Dosing Cell Engineering Diagrams and Schematics

Engineering diagrams for a gas dosing cell designed to accommodate a porous Si chip prepared using the Standard etch cell. The design incorporates a slot for an optical window (standard glass or quartz microscope slide) that can be used to monitor the porous Si sample by optical reflectivity, fluorescence, or light diffraction techniques.

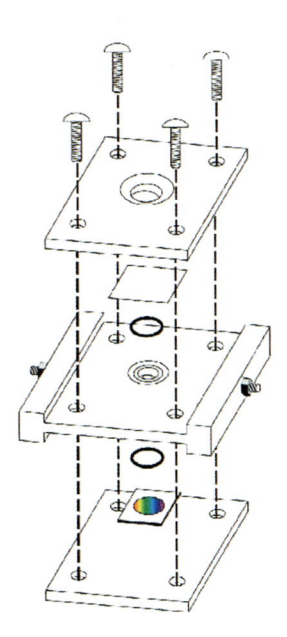

Appendix A3. Gas Dosing Cell Engineering Diagrams and Schematics

Gas Dosing Cell Top Piece

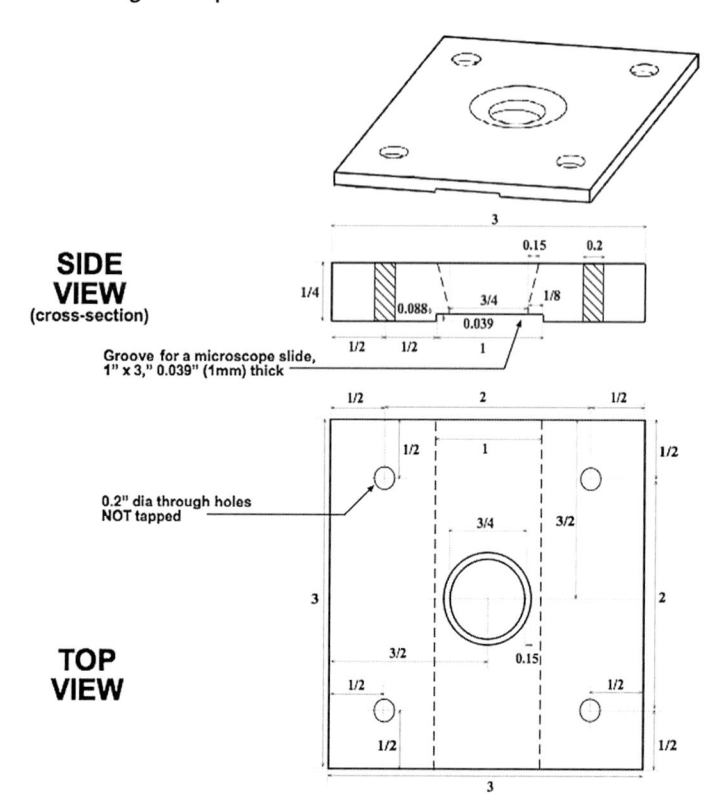

SIDE VIEW (cross-section)

Groove for a microscope slide, 1" x 3," 0.039" (1mm) thick

TOP VIEW

0.2" dia through holes NOT tapped

Appendix A3. Gas Dosing Cell Engineering Diagrams and Schematics

Gas Dosing Cell Middle Piece

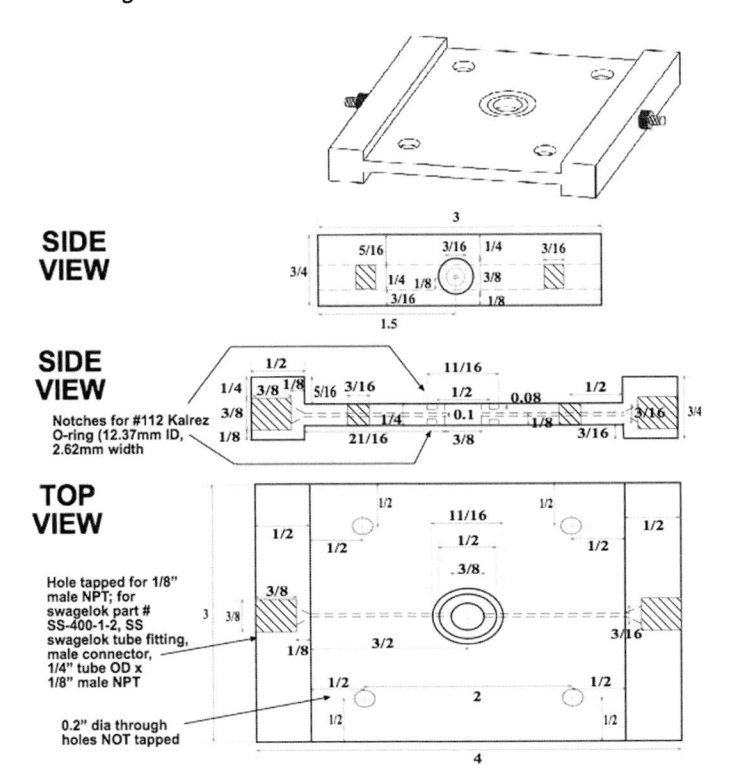

SIDE VIEW

SIDE VIEW
Notches for #112 Kalrez O-ring (12.37mm ID, 2.62mm width)

TOP VIEW

Hole tapped for 1/8" male NPT; for swagelok part # SS-400-1-2, SS swagelok tube fitting, male connector, 1/4" tube OD x 1/8" male NPT

0.2" dia through holes NOT tapped

Appendix A3. Gas Dosing Cell Engineering Diagrams and Schematics

Gas Dosing Cell Bottom Piece

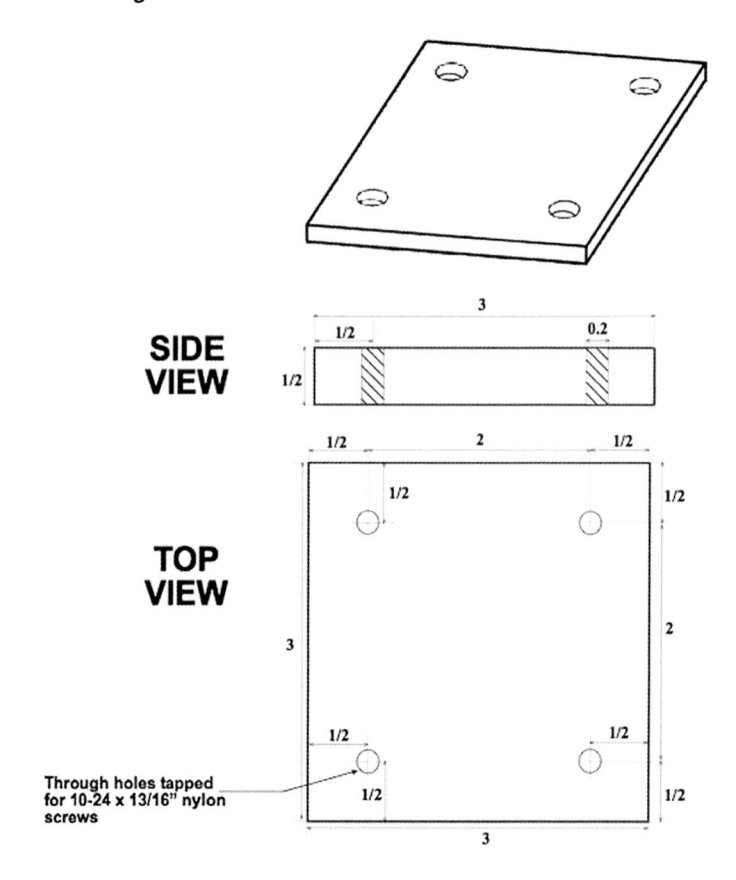

SIDE VIEW

TOP VIEW

Through holes tapped for 10-24 x 13/16" nylon screws

Index

a
additives 3, 198
annealing 193
applications
– biosensors 189, 213
– biotechnology 58, 189
– Bragg mirrors 165
– drug delivery 12, 189
– film displays 83
– microcavity resonators 165
– microelectromechanical systems (MEMS) 112
– optical filters 102
– rainbow chips 106
atomic force microscopy (AFM) 12, 105f., 139
– asymmetric electrode etching 105f.
– porous silicon from p-type wafer 54

b
band
– conduction 21f., 25
– energy 24f.
– gap 21, 176
– stop 84
– valence 21f., 25
BET (Brunauer–Emmett–Teller) adsorption method, see characterization methods
biocompatibility 189, 199, 212f.
biomolecule conjugation, see conjugation of biomolecules
bonds
– Si–X 2, 189f., 200f.
Bragg stack, see photonic crystals
Bruggeman model, see characterization methods
byproducts 199

c
CCD (charge coupled device) 22
– spectrometer 140ff.
cell
– fabrication materials 44f.
– gas dosing 241ff.
– large etch 229, 231f.
– power supply 44f., 65f.
– safety precautions 48ff.
– small etch 229, 231f.
– standard etch 105, 179, 206, 229f.
– three-electrode 6f.
– two-electrode 5f., 43
characterization methods
– Barret–Joyner–Halenda (BJH) adsorption method 167, 170
– Broekhof–de Boer (BdB) adsorption method 167
– Bruggeman model 149, 154ff.
– Brunauer–Emmett–Teller (BET) adsorption method 12f., 167f., 170, 196
– fast Fourier Transform (FFT) 151ff.
– Fourier transform infrared (FTIR) 176ff.
– infrared (IR) spectroscopy 176ff.
– optical reflectance measurements 133, 139f., 197
– Raman spectroscopy 176
– reflectometric interference Fourier transform spectroscopy (RIFTS) 143, 150f.
– scanning electron microscopy (SEM) 138
– spectroscopic liquid infiltration (SLIM) method 134f., 143, 149, 151, 154ff.
– transmission electron microscopy (TEM) 138

Porous Silicon in Practice: Preparation, Characterization and Applications, First Edition.
Michael J. Sailor.
© 2012 Wiley-VCH Verlag GmbH & Co. KGaA. Published 2012 by Wiley-VCH Verlag GmbH & Co. KGaA.

charge
- carriers   21f., 36, 175
- carrier recombination   24, 175
- transport   26
- tunneling   57
cleaning silicon wafers   51
conductivity   21, 24, 57
conjugation of biomolecules
- attachment of PEG   212f.
- biomodification of hydrocarbonized porous silicon   213f.
- carbodiimide coupling reagents   211f.
- silanol-based coupling   215ff.
contact angle   203, 205, 215
cracks   59, 105, 189, 198
crystal face selectivity   18
current
- blocked   29, 63
- corrosion   17
- density   16, 19, 27, 48, 58f., 92
- leakage   27
- photo-   29
current–potential curve   7, 10
current
- reverse saturation   27, 29
- time-programmable   78f., 81f.
- –time waveform   100ff.
current–voltage curve   26
current source   46f.
- voltage-programmable   79, 81f.
crystalline silicon, see silicon

d

DAC (digital to analog converter)   78f.
diode
- ideal law   27, 29
- photo-   22
- Schottky   36f.
dissolution
- electrochemical   13
- silicon   2
- silicon oxide in basic media   3
- silicon oxide in HF solutions   3, 121
DMM (digital multimeter)   38, 57f.
dopants   22f., 30
- density   92
- segregation   59
doped
- highly   13f., 79, 85, 88, 94, 107f., 122ff.
- low-   79
- n-type   13f., 19, 23, 25, 37ff.
- p-type   8, 13, 23, 25, 37ff.

e

electric field distribution   18
electrochemical
- anaerobic   206
- anodization   68
- corrosion reaction   3, 6, 8f., 16, 27
electrochemical etching   5, 8, 16, 57
- catalyzed   73
- metal-assisted stain etch   73f, 74ff.
- photoetching   111f.
- power supply   78ff.
- programmed   78ff.
- reaction stoichiometry   17, 56
- reliability   72
- reproducibility   61, 72
- side   105f.
- stain   68, 193
- time resolution   79f.
- waveform superposition method   104
electrochemical
- grafting   206f.
electrocorrosion reaction, see electrochemical corrosion reaction
electrode
- asymmetric configuration   104ff.
- counter-   6, 8, 10, 26, 44, 59, 69, 104ff.
- gate   22
- reference   7
- saturated calomel   7
- working   6, 8, 59
electron   21ff.
- –hole pairs   28, 59
- transfer kinetics   9, 192
electropolishing   7ff., 106
enthalpies   2
equilibrium constant   24
etch cell, see cell
etching, see electrochemical etching

f

Fabry–Pérot fringes   91, 101, 139, 143, 147, 150, 154f.
- layer   145ff.
Faraday's constant   56f.
fast Fourier Transform (FFT), see characterization methods
Fermi
- energy   22
- level   21
flakes   59, 198
fluorescence spectra   174
freestanding films, see lift-off films

g

gas adsorption measurement 12f., 167f., 170, 196

h

half-reactions 5ff.
HF (hydrofluoric acid) 3
– electrolyte 5, 51
– safety precautions 48f., 235ff.
hole 8, 10, 21ff.
– conduction 22
– valence band 16f., 19
hot-probe method 36
hydrosilylation 200, 203f., 217f.
– thermal 200, 203f.
– microwave-assisted 204f.

i

impurity 22f., 51
index of refraction 156ff.
infrared (IR) spectroscopy, see characterization methods
interface
– air/porous silicon 91, 103, 144
– bio- 207
– passive 175
– pore/silicon 83
– semiconductor/electrolyte 25
– silicon/bulk silicon 103, 120, 144
– silicon/crystalline silicon 10, 13, 17, 91
– silicon/electrolyte 38, 63, 79
– silicon/metal 39
– silicon/porous silicon 103, 120, 144
– silicon/solution 104
interfacial transport 27
ionization constant 4

k

Kelvin equation 168
Kohlrausch–Williams–Watts stretched exponential model 174f.

l

layer
– double 71, 79, 84, 87, 150, 162ff.
– multi- 71, 89f., 150
– refractive index 83, 88
– single 43, 101, 151, 162, 165
– spatially modulated 77ff.
– transparent 144
lifetime
– emission 174f.
– nonexponential excited state 173f.

lift-off films 10, 17, 121ff.
– hydrogen-terminated 126
liquid junctions
– current–voltage curve at n-type Si 28f.
– current–voltage curve at p-type Si 26f.
– energetics at n-type Si 28
– p-type Si 26f.
luminescence
– bands 172
– macroporous silicon 60ff.

m

macropores 12ff.
masks, see patterning
mass transfer effects 59
mesopores 12ff.
– double layer by programmed electrochemical etch 85ff.
– preparation from a p^{++}-type wafer 53ff.
metal-assisted etch, to form nanowires 73f.
metallization reactions 218f.
microparticles 120
micropores 12f., 53
Miller indices 18, 20
modulation
– nanostructures 111
– porosity 83, 85, 95, 111
– refractive index 83f., 88

n

nanocrystallites 59
nanofibers 209f.
nanoparticles 73, 121, 196
– core/shell (Si/SiO_2) 126f.
nanopores 123f.
nanowires 57, 73, 73ff.
– optical absorption spectra 170
– quantum confinement 170
– vapor–liquid–solid (VLS) growth 74
Newton's rings 154
nucleophile 5, 17, 207
nucleophilic attack 18
Nyquist–Shannon theorem 80

o

ohmic contact 38ff.
– mechanical abrasion 40
– metal evaporation 39
open circuit potential (OCP) 8

oxidation
- aqueous solution 193ff.
- dimethyl sulfoxide (DMSO) 192, 195ff.
- four-electron electrochemical 8ff.
- gas-phase oxidants 190
- organic species 195
- rapid thermal (RTO) 192, 196
- two-electron electrochemical 8ff.
- thermal 164, 167, 192

p

patterning 77, 108ff.
- after etching approach 110
- before etching approach 110
- physical masks 108ff.
- virtual masks 108, 112f.
peeling 105
photoelectrochemical etching
- back side illumination 69f., 111
- front side illumination 66ff.
- n-type silicon wafer 65f.
photoexcitation 22, 59, 63
photoluminescence (PL) 59, 120, 127, 170ff.
- instrumentation 173
- mechanisms 171f.
- steady-state spectra 170ff.
- time-resolved spectra 173
- tunability 171
photonic devices 71, 80, 83ff.
- Bragg reflector 96f.
- Bragg stacks 83f., 94ff.
- films 119ff.
- microcavities 94f., 165
- multilayered 1-D 89
- multi-line spectral filters 94f., 100f.
- reflectivity spectrum 101, 103f.
photoresist 109f.
polarization memory 174
pore
- bottoms 13
- branching 19
- crystallographic 18f.
- current 19
- diameter 57, 77, 87, 104, 111
- expansion chemistry 197
- formation 13f.
- modulation 83
- morphologies 13f., 18, 64, 67
- size 12, 58
- texture 14
porosity 4, 11, 87, 104ff.
- average 88

- closed 12
- –depth profile 103
- gradient 77, 85f., 104ff.
- gravimetric determination 134ff.
- high 105
- open 11, 155
- total 12
porous glass membrane 7

q

quantum confinement effects 18, 57, 59, 127, 170, 174, 176

r

recombination, *see* charge carrier
reflectivity spectrum 101, 103f., 133
reflectometric interference Fourier transform spectroscopy (RIFTS), *see* characterization methods
reproducibility
- electrochemical etching 57, 69
- thermal oxidation 189
resistivity
- four-point experiment 30ff.
- rugate filter 91
- silicon wafer 29f., 31ff.
rugate filter 83ff.
- color-graded 107f.
- current–porosity 94
- refractive index–porosity 92f.
- spectral band width 92ff.
rugate spectral peak wavelength 88

s

scanning electron microscopy (SEM) 12, 138f.
- branched macroporous morphology 68
- cross-sectional 19, 68, 71, 138, 157
- double layer 84
- effect of dopant on pore texture 13
- effect of surface cleaning on pore morphology 60f.
- high resolution (HR-SEM) 12
- n-type porous silicon 71
- photonic crystal particle 111
- plan-view imaging 139
- silicon nanowires 73
- stain-etched silicon powders 71
Schlenk line 200ff.
Schottky barrier 36
silicon
- band structure 20
- crystalline 3, 19, 191

– hydrides   4f.
– intrinsic   23
– orientation   19
silicon oxide (SiO$_2$)
– insolubility   4
– water-soluble forms   4
silicon
– powders   68
– unit cell   19
silicon wafer   10, 28ff.
– cleaning   51f., 60f.
– cleaving   34f.
– crystal orientation   30f.
– dopant type   30f.
– refractive index   83, 88, 92f.
– resistivity   29f., 31ff.
– single-crystal   104
– thickness   31, 87, 104f.
– thickness gradient   105
skeleton
– index of refraction   156ff.
– index on porosity   158f.
space-charge region   19f.
spectral barcodes   110, 120
spectroscopic liquid infiltration (SLIM), *see* characterization methods
surface
– area   59
– crust layer   60

– hydrophilic   204, 215
– hydrophobic   204
– radicals   214f.
– -to-volume ratio   175
– traps   175

t

thermal
– carbonization   208ff.
– degradation   208f.
– oxidation   164, 167, 192
thermally carbonized porous silicon (TCPSi)   208f.
thickness   31, 87, 104f., 134ff.
– effective optical (EOT)   147
– gradient   105, 138
– gravimetric determination   134ff.
transmission electron microscopy (TEM), *see* characterization methods
transparent porous silicon layer   144

u

ultrasonication   120ff.
– cleaner   125f.
– fracture   119, 122f.
unit cell of silicon   19
UV light   59, 67, 112, 173

w

water splitting potential   9